# Author Experience

Bridging the gap between people and technology in content management

Rick Yagodich

Foreword by Noz Urbina

# Author Experience
*Bridging the gap between people and technology in content management*
Copyright © 2014 Rick Yagodich

All rights reserved. No part of this book may be reproduced or transmitted in any form or by any means without the prior written permission of the copyright holder, except that brief quotations may be used with attribution, for example in a review or on social media.

## Credits

| | |
|---|---|
| **Series Producer and Editor:** | Scott Abel |
| **Series Editor:** | Laura Creekmore |
| **Series Cover Designer:** | Marc Posch |
| **Copy Editor:** | Ray Johnston |
| **Indexer:** | Cheryl Landes |
| **Publisher:** | Richard Hamilton |

## Image credits

- Illustrations in Figure 1 and Figure 5.4 copyright © 2012–2013 Chris Shipton, used by permission.
- Portrait in Figure 4.1 and Figure 7.5 copyright © 2013, Reyner Media, CC-BY-2.0.
- Cover design and The Content Wrangler logo copyright © 2013 The Content Wrangler.
- XML Press logo copyright © 2008 XML Press.

## Disclaimer

The information in this book is provided on an "as is" basis, without warranty. While every effort has been taken by the authors and XML Press in the preparation of this book, the authors and XML Press shall have neither liability nor responsibility to any person or entity with respect to any loss or damages arising from the information contained herein.

This book contains links to third-party websites that are not under the control of the authors or XML Press. The authors and XML Press are not responsible for the content of any linked site. Inclusion of a link in this book does not imply that the authors or XML Press endorse or accept any responsibility for the content of that third-party site.

## Trademarks

XML Press and the XML Press logo are trademarks of XML Press.

All terms mentioned in this book that are known to be trademarks or service marks have been capitalized as appropriate. Use of a term in this book should not be regarded as affecting the validity of any trademark or service mark.

XML Press
Laguna Hills, California
http://xmlpress.net

First Edition
978-1-937434-42-7
978-1-937434-43-4

# Table of Contents

Foreword ................................................................. v
Preface .................................................................. vii
**I. Author Experience Basics** ........................................... 1
   1. Introduction ...................................................... 3
   2. What is Author Experience? ........................................ 7
      Which author? ................................................. 7
      The role of content management ................................ 8
      A definition of author experience ............................. 10
   3. The Current State of Author Experience ............................ 13
      Messy system, messy content ................................... 13
      Purchasing a platform ......................................... 15
      Weighing author experience against user experience ............ 19
      Taking the lead in improving AX ............................... 20
   4. The Challenges to Good Author Experience .......................... 21
      Knowing how to communicate .................................... 21
      The communication goal ........................................ 23
      The technologists' paradigms .................................. 25
      Content coupling .............................................. 30
      Workflow that... doesn't ...................................... 33
      Content ownership ............................................. 37
      Mental models ................................................. 40
      Metadata ...................................................... 46
**II. Practical Author Experience** ...................................... 51
   5. Hierarchy of Author Experience Needs .............................. 53
      Fit-for-purpose language ...................................... 54
      Content accessibility ......................................... 62
      Associative structured content ................................ 66
      Rules-based presentation ...................................... 75
      Content management tools ...................................... 81
      Self-aware content ............................................ 86
   6. Conducting an Author Experience Audit ............................. 93
      Goals of the AX audit ......................................... 93
      Tools of the trade ............................................ 95
      Technical business analysis ................................... 96
      Content design and governance ................................. 98
      Author experience design ...................................... 101
      Information architecture ...................................... 109
      Auditing ...................................................... 111
      Measuring author experience quality ........................... 112
   7. Author Experience Design Patterns ................................. 117
      UI label management ........................................... 117
      Integrated label use .......................................... 120
      Label structure deployment .................................... 122

## Table of Contents

    Content consoles .................................................. 124
    Narrative ............................................................ 127
    WYSISMUC ........................................................ 129
    Selection by date ................................................. 132
    Referrer / referee links ........................................ 134
    Stored or searched references ............................... 136

**III. The Future of Author Experience .................... 139**
  8. Moving Forward with Author Experience ............. 141
    A new name for old rope? .................................... 141
    The customers are more important ........................ 142
    It's too much work ............................................. 142
    CMS vendors ..................................................... 143
    Conflict of author interest .................................... 143
    Silos (Who owns content?) ................................... 144
    The Agile battlefield ........................................... 145
  9. In Conclusion .................................................. 147
    The evolution of content management ................... 147
    Moving forward .................................................. 150
Glossary .................................................................. 153
Index ...................................................................... 157

# *Foreword*

Rick Yagodich has set himself a major challenge: bring the market a message that it really needs, but doesn't really want to hear. Or at least, it doesn't yet realize it needs to hear it. He has taken on the daunting task of telling those who fund, select, implement, and design configurations for content management systems to do something quite difficult: think of other people first.

If we're brutally honest, many of us will find it intuitively true that in business we often put ourselves in the middle of the question being asked. Our roles, our backgrounds, our ways of thinking, and our language color our understanding of business challenges and, especially, their solutions. As the saying goes, when you have a hammer, everything looks like a nail. But our hammers are simply not good enough. With the diversity of challenges intrinsic to today's content delivery, they're no longer fit for purpose (if they ever were).

We're in an industry where change rains down so hard and fast that looking several steps ahead is the only way to avoid being swept away and drowned. We haven't managed to get everyone up to speed on fundamentals like structure and semantics, much less domain models, adaptive content, linked data, or augmented reality. Nevertheless, because of the nature of this market, we have no choice but to keep building up our platforms while the change-storm rages.

So far, we've been using our hammers as best we can to build a protective shelter, but we've arguably had to go about it in a backwards way.

We started at visual web design and SEO, and worked inward. It was like building the roof to keep us dry before building the house or laying the foundations. Inevitably, the results were fragile and imperfect, at best. Many of us got wet – or positively soaked. In recent years, information architecture and user experience have become recognized pillars to hold the roof up. That was good progress, but it still left us exposed. Now we're finally realizing we need the rest of house to keep us safe, productive, and dry for the years to come; content strategy, semantic structures, smarter content models, and omnichannel thinking are all now gaining mass-market understanding and will serve us well.

What we are still lacking to give our content agility – to really keep the house strong over time – is attention to the foundation it is all built on: authors. This is where author experience comes in. At long last, but not too late, organizations are waking up to the fact that author experience must be addressed to optimize our solutions. We've lacked a champion of that indispensable class of users who create all this "stuff" for us in the first place. Rick has stepped up. This book describes and illuminates a discipline that unites the author experience on one end with end-user experience on the other. He describes how applying this discipline can generate benefits for users of both types, and the organizations with which they're engaging.

I call *Author Experience* a "second generation" book about content solutions. While reading it you'll see that it implicitly says, "For 'Intro to's and '101's on the other content concepts there are already books out there for you. To strengthen our house we have new ground to break, new work to do, and new tools to develop."

At some point you need to get beyond the basics and reach new levels of sophistication, understanding, and refinement. If your strategy in the content world is "Just do the basics and plug the worst leaks," you'll quickly find yourself neck deep in issues. The market's evolving too fast for a keep-up strategy. This book will bring you the knowledge you need to resolve issues you hadn't fully recognized you had, yet. Read it, and if needed, read it again. Read the footnotes. Take it all in. When you have, you'll see its real message is simple: author experience is vitally important, it's complicated, and it's broken; so now let's get started on fixing it.

Noz Urbina

# *Preface*

For many years, I considered myself an information architect; not one content to do wireframes, but someone who insists on looking at the structure and interaction of elements within an information system. This was predominantly within the web environment.

Then, I had a few projects relating to the customization of a particular web content management system (CMS) that demonstrated what I could think of only as horrors: inconsistencies in how elements of functionality worked out of the box, technical paradigms surfaced to the author, customization that was even more piecemeal than the original system. It was obvious that with such a messy interface, there was no way anyone could create and maintain the kind of quality content needed for a good end-user experience.

Figure 1 – Maestro[1]

Figure 1, "Maestro," is a perfect representation of the state of the CMS, or was that CMmess?

From this, I coined the term *author experience* (AX) in 2010, wrote a short piece about it (*The experience alphabet: AX comes before UX*[2]), and started to establish working principles. This was not easy; every

---

[1] By Chris Shipton, http://chrisshipton.co.uk/, used with the artist's permission

[2] http://www.cognifide.com/blogs/ux/the-experience-alphabet-ax-comes-before-ux/

project is different, and every client has its own needs. But this work reinforced just how important it is to address the challenge of authoring within digital systems.

It did not take long to see that these problems are not limited to web CMSes. Technical communication CMSes have solved some of the issues, but at the cost of making other aspects of authoring more difficult. While the term author experience and the working principles that author experience entails expand beyond the web, my background does mean that I can accidentally drift in that direction when discussing the subject. So if at times I come across as web-centric, please forgive the lapse.[3]

In business, there is a growing understanding that marketing material and support material need to overlap and interact. But we are not limited to these two types of material; we need information that can move across all aspects of a business without manual duplication and its associated risks. Avoiding the disruptions associated with poor content management provides a clear business value. Indeed, it is the best route to justifying author experience.

While I struggled with a way to better promote author experience, a conversation with Scott Abel led to the suggestion that this book would be a good addition to his Content Wrangler series on Content Strategy. And so it happened.

This book looks at author experience from a series of angles:

- A definition of the scope: the business value of author experience and the aspects of existing systems that make content management so difficult.
- A practical perspective on author experience: a staggered approach to authoring functionality that improves the experience, and advice for putting author experience principles into practice.
- A look at challenges we can expect to encounter before author experience becomes a generally accepted field.

It is aimed at a cross-section of people involved in content management, from those with business oversight of the function, through those who specify and those who use content management systems, to those who implement the underlying technology. For those who want only a high-level understanding, I recommend Part I, "Author Experience Basics," and Chapter 9, *In Conclusion*. For everyone else, all chapters are relevant.

---

[3] That said, isn't everything in the world now moving to the web?

While I expect readers to have some understanding of content management, I do not expect my audience to have any background in author experience. As such, I will build up a common frame of reference.

> **Asides**
>
> To include examples without deviating from the flow of the book, I include most references to my personal experience in the form of asides like this one.

## Disclaimer & Explanation

If we were to meet, you might wonder how I could be an authority on any aspect of communication. Whereas someone like Bill Gates might be renowned for being socially awkward, I manage to personify socially inept. I am – through and through – an Aspie. (That's Asperger Syndrome,[4] in case the terminology is unfamiliar.)

This means that while most people communicate 60% through body language, 30% through tone, and only 10% through the words they use, that process does not work for me. I am blind to body language directed at me, and much tonal communication – especially when subtle – fails to add to what I perceive from the words.

Similarly, the majority of context within most people's communications is assumed or implied. I find it quite fascinating that meaningful conversations can be held when no baseline has been established to give meaning to the discussion.

You could be forgiven for expecting that this state of affairs – it is not a disorder, despite some classifications – would interfere with all types of communication. From my perspective, it is an advantage. Because I do not see information exchange in the same way most people do, and because I do not make assumptions about the assumptions my counterparts make about context, I have much experience deconstructing and analyzing the elements that make up the message.

Until computers can emote, they will exchange information only as I do: through the words and images used, with but a smattering of tone.

Author Experience is about mapping the mechanics of communication to a logical and contextually usable set of interfaces, which is how I must engage in my day-to-day communications.

---

[4] See Asperger Syndrome [http://en.wikipedia.org/wiki/Asperger_syndrome] on Wikipedia.

## Acknowledgements

I would like to thank a few people whose input, ideas and feedback have led to specific improvements in this work (or just made it possible): Scott Abel, Sally Bagshaw, James Deer, Jeff Eaton, Max Johns, Richard Hamilton, Elizabeth McGuane, and Noz Urbina.

Maybe I should also thank certain clients and CMS developers for showing me just how bad it is out there… But I wouldn't want to embarrass anyone.

Rick Yagodich,
London, August 2014

# Author Experience Basics

# CHAPTER 1
## *Introduction*

Who isn't familiar with this story? Your company needs a new system to manage content. The system is designed, implemented, and delivered. Then, the people who must use the system discover that it is as bad as, or worse than, the old system. You wonder why anyone bothered.

Author experience (AX) focuses on the value of managing the communication process effectively and efficiently. It deals with this process from the point of view of those who create and manage content. It also aims to avoid the pitfalls that all too often lead to systems that are worse than what they replaced.

While author experience largely means user experience within a content management environment, it is concerned with issues – creating and managing the communicated message – that often suffer from a myopic focus on end-user experience. Author experience looks to remove the inconsistencies and cognitive hurdles that make it harder than necessary to work with content management systems.

While the practical aspects combine existing disciplines, author experience is more than a new name for old processes.

## The communication process

Let's start by looking at the content communication process. This process flows from communicator to audience (Figure 1.1). Bidirectional communication is the same, with the direction repeatedly flipped.

Figure 1.1 – Basic communication

Viewed this way, it is pretty simple. But looking closer, we see there are three parts: input, storage, and output (Figure 1.2).

Figure 1.2 – Input, storage, output communication

When we have a single channel, these parts resemble each other. Input is the creation of content, storage is a delay, and output is the presentation of that content in the same form as the input. However, when we want our output to be presented across multiple channels, we must disassociate the storage paradigms from the output. We need a layer that translates between the storage and the various channels (Figure 1.3).

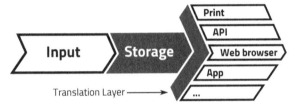

Figure 1.3 – Adaptive output communication

Once we disassociate the output, we can optimize the storage. But this optimized storage model may not be suitable for those creating the content. We need a further layer of translation so that the input logic and paradigms make sense to authors, rather than simply mirroring the storage model (Figure 1.4).

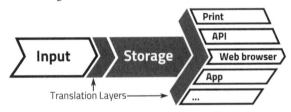

Figure 1.4 – Authoring adaptive communications

Author experience is concerned with ensuring that this translation layer provides a natural management interface, suitable for authors, that connects to a structured storage environment suitable for multi-channel adaptive output.

The field of author experience seeks to redress a balance, to give a name to an area of content system design that is losing out because (end) user experience is the flavor of the ~~month~~ ~~year~~ decade,[1] claiming more than its fair share of budget and effort. Where technology was originally the primary driving force, the balance has shifted now to the communication, but is still lopsided in favor of the recipient. However, if the tools are not fit for the purpose of creating and managing content, how can we ever create that optimal end user experience?

---

[1] I do not deny that user experience was sorely neglected for far too long, so I understand why there has been such a backlash in its favor. But there is such a thing as too much of a good thing.

# CHAPTER 2
# *What is Author Experience?*

To define author experience (AX), we first need a common understanding of *author*, both as a person and an environment. We also need to look at the role of *content management systems (CMS)*.

## Which author?

The word *author* is used within different branches of the content industry to identify different groups of people. So that we all have a common frame of reference, here is how I use the term in this book.

### *The author environment*

The term *author environment* identifies the interface to a system where information is created and managed, as opposed to the medium to which the content is published.[1] While the term derives from enterprise web content management, it is equally applicable to other platforms. When referring to the author environment, I will always use that specific term, to differentiate it from authors as people.

### *The human author*

Many people, with many roles, use the author environment of content management systems. In this book, the term *author* encompasses all of these roles, not just the people who write.

> An **author** is anyone who interacts with the CMS.

This definition includes anyone involved in the specification, structure, design, writing, editing, approval, and maintenance of content. It covers, but is not limited to, the following roles:

- **Information architects:** define the underlying content structure
- **Content strategists:** establish governance principles and workflow
- **Subject matter experts (SMEs):** create raw material
- **Writers:** craft raw material into readable and digestible content
- **Business stakeholders:** approve content
- **Editors:** pull the content together
- **Translation managers:** maintain localized versions of content
- **Designers:** identify the optimal display for content

---

[1] In enterprise web content management, the address of the author environment is often in the form author.domain.com

## The role of content management

Restricting author experience to the web detracts from the subject's scope. While this may make it easier to grasp initially, it severely hobbles the author's abilities within the system. The expansion of content delivery beyond the web is what makes author experience so important.

### *The scope of content management*

Because content management and content management systems are core to author experience, we need to understand the full scope of what we are talking about.

Any system that enables information to be created and stored for later retrieval is at least a content storage system. By "stored for later retrieval," I mean communication with an audience, which may or may not be defined, where transmission is suspended for an indeterminate period. If the communication is not a real-time conversation,[2] it probably involves content storage.

To qualify as content management, the content storage needs to provide:

- Some type of compartmentalizing structure, to break down the information into logical blocks, and
- The ability to edit information in the content store.

It is easy to see that an application for managing a web site qualifies as a content management system.[3] Likewise, the various business systems that handle product inventories, purchasing, sales, and finance are all types of content management systems. The inputs and outputs may not be human interfaces, but this does not detract from their function.

Taking this a step further, we realize that just about any computer system qualifies as a content management system. Even something as archaic as a filing cabinet is a content management system. So is a ring-binder. But a notebook is not: it lacks a function – page replacement – required to edit the material within it.

---

[2] There is the case of extreme long distance communication – think interplanetary – that fails to be "real-time," though this is due to the limitations of the carrier medium; we just can't hack the speed of light yet.

[3] This is not an endorsement of any such system, only an admission that they fall within the concept's scope.

## Why content management?

Knowing what a content management system is, let us now ask: what is its purpose?[4]

Depending on who you ask, the answer to the question "What is the purpose of a CMS?" will vary wildly (see Table 2.1 for a few of these answers), but the answers have one common problem.

Table 2.1 – Reasons for a content management system

| | |
|---|---|
| Web CMS vendor | A CMS is the way you store information to present on a web site or other platform. |
| "Experience" WCMS vendor | A CMS is used to store and present content to enhance your sales and online conversions. |
| CMS integrator/ content modeler | A CMS is used to make sure your content follows standardized patterns so it is more easily reused. |
| Web designer | A CMS makes content presentation more consistent (and gets in the way of my artistic efforts). |
| Marketing department 1 | If it's linked to our sales database, a CMS enables us to cross- and up-sell better to our customers; if not, it just gets in the way. |
| Marketing department 1 | We can quickly create content without having to learn any pesky web skills. |
| User assistance people | The CMS gives us a pool of information that we can reuse. |
| Compliance officer | A CMS provides an audit trail of content versions, to satisfy regulatory requirements. |
| Content manager | The CMS makes us jump through hoops to maintain our content. |

The problem is that none of them answers the actual question. They all address use or features or functionality, but not purpose.

---

[4] Many have asked this question. Unfortunately, too often the question arises after a system implementation, when those tasked with managing content have found that the system hampers more than helps.

## The purpose of content management

*Purpose* relates to a reason for existing – a fundamental aim related to being: the benefit it will bring someone, rather than its functionality. Purpose serves a value proposition.

So, what purpose does a content management system serve?

At its functional heart, a CMS enables the author to enter content, manipulate it, and provide for it to be extracted, in some form, for presentation or some other use. Why would we want to do that?

The answer is easily overlooked: it enables communication.

All this stuff we call *content* within a CMS is just information.[5] The purpose of information is to be exchanged. The exchange of information is communication. Therefore, a CMS is a tool that serves communication.

If a CMS is any system that enables management of information being used in non-real-time communication,[6] then the purpose of a CMS is to allow us – people – to engage in these communications with partial automation. Thus, the real reason to have or use a CMS is to facilitate a human process – managing the information within a non-real-time communication. And taking the full scope of that into account, we see that the purpose covers the entire lifecycle of such content.

So, succinctly:

> The **purpose** of a content management system is to **facilitate the human process** of managing the communication content lifecycle from creation, through use, to archiving.

## A definition of author experience

With the purpose of content management systems clear, we can look to the definition of author experience.

---

[5] The terms *content* and *information* have been so abused by the industry that I need to clarify how I differentiate them. *Information*, for me, is anything that has meaning. *Content* is the subset of information that has human interaction. They are almost-synonyms.

[6] I am aware that the term *real time* is being used increasingly for digital communication. Such exchanges are rapid, but few are truly real time. One process takes, encodes, and stores the elements of the communication, then another propagates them. The point is that the propagation *can be* delayed without the system breaking. Of course, there are edge cases: a video conference running through a digital system has a foot in each camp.

Some would say it is just user experience within a CMS authoring environment. To some extent, that is true, but it fails to clarify how the two disciplines differ. User experience aims to make using something easier and more enjoyable, pleasant, or engaging. And while we might want that also to be true of author experience, those emotional benefits are secondary.

Content authoring is not – for most people – something that qualifies as fun. It is a task at best, often a chore or nuisance. It is a business activity.[7] So the most important facet is not that it be enjoyable, but that it be efficient, logical, and appropriate to the task.

With this in mind, we see the difference in focus between AX and UX. The best single word to describe what we are aiming for with author experience is appropriateness. The environment should suit the function of the task. We want the right range of functionality, but only to expose appropriate functionality to the author's context.

As such:

> Author experience, as a discipline, is the provision of **contextually appropriate functionality** within a content management environment.

There are many ways we can look at this goal of contextually appropriate functionality; we will cover several in Part II, "Practical Author Experience."

## *Why now?*

As a quick aside, here is a little context: how business survived two-plus decades of the digital age without worrying about author experience.

For many years, digital systems have been effectively one dimensional. A single content management environment served a single output. There was no reuse. It served the authors' mindset to match the appearance of the authoring environment to the way the content would be used.

As digital media spread into more corners of our lives and the methods through which we consume information diversify, so the content needs to adapt to more environments, more forms.

---

[7] If you are of the mind that content authoring for a personal project is not a business activity, consider why you are doing it. What are you aiming to get from it? What is the effort going to allow you to exchange that content for?

## What is Author Experience?

Authoring is becoming more abstract and more complex, making life significantly more difficult for authors and increasing the risk of errors and inconsistencies. These trends will continue unless we focus on new paradigms that enable us to manage multi-channel content in a single environment without undue pain. As the Editorial Product Manager at National Public Radio (NPR), Matt Thompson, wrote in 2011:[8]

> "Your average news organization CMS used to require weeks of training to acclimate new users…. Beautiful software, even for back-end users, is becoming an expectation. We're moving in this direction because we now understand that better content management systems foster better content."
>
> —Matt Thompson, National Public Radio

---

[8] http://www.poynter.org/how-tos/digital-strategies/134791/4-ways-content-management-systems-are-evolving-why-it-matters-to-journalists/

# CHAPTER 3
# *The Current State of Author Experience*

The idea of author experience is often dismissed by people with budgets and those who design and build CMSes. They see more important places to invest, better opportunities for quick-win returns. Author experience demands a long view; it suffers in the face of short-term investment attrition. We must, therefore, demonstrate the loss, and the risks, incurred by ignoring this opportunity.

## Messy system, messy content

So far, we have three basic pillars:

- Authors constitute anyone who interacts with our content management system.
- The purpose of a CMS is to facilitate the communications process for authors.
- Author experience is the discipline of implementing the part of a CMS that enables it to fulfill its purpose.

It all sounds straightforward. Figure out what authors need to do their jobs better and accomplish their tasks with the least pain and disruption, then implement that. Right?

Yeah, right!

### *The state of the CMmess*

Considering only those who classify themselves as CMS providers, there are hundreds of players. Some offer enterprise implementations with significant license and consulting fees. Some offer usage-specific, open-source platforms that anyone can set up in a matter of minutes. Most of the rest offer something in between. Then there are players who offer custom implementations, often derived from existing platforms.

Commercial CMS vendors win contracts by offering a set of features that looks good in a presentation deck and impresses the client's marketing, IT, and purchasing departments. It doesn't matter how well those features work because vendors know that clients will want the system customized to their needs; anything out-of-the-box will be modified.

> **A features list, without integration**
>
> A couple of years ago, while solving some issues for a client, I ran into the following problem.
>
> The platform, for which they had paid seven figures, provided an element that let you associate a block of text and an image. It also had a localization model that enabled content structures and elements to stay synchronized.
>
> Great, you might think. We need both features. Except, when you edited a localized version of the text, the content synchronization was lost *at the level of the entire element!* Change the text, and the image was no longer synchronized with the master content instance. Any change to the image now needed to be replicated to 26 locales.
>
> This platform ticked both feature boxes, but in practice, it failed completely.

Open source platforms are built by people who are technologically competent; they understand the workings of their systems. And because they are familiar with the technology and storage/retrieval paradigms, they do not find it onerous to manage their content using those paradigms; the model makes sense to them. However, the non-technical author is left out in the cold.

Custom implementations fare somewhat better. Because they are usually built for one client's specific task, they stand a decent chance of doing that task well. But rarely do any improvements make it out into the wider world; the value is not shared.

Content management, generally, is a mess. Almost no attention is paid to the people who use content management systems. They are left to decipher interfaces and paradigms that bear no relation to the way they think about the content they manage. How the system presents content is sexier than how well it enables authors to manage content; the end-user experience is given more weight than the author experience.

## *Exposing the value*

These problems engender frustration and drain resources. If you are trying to solve a content problem and the CMS forces you to think in a way that does not make sense to you, you cannot be productive. Frustration results in a lack of attention. If something that should be handled in a single process requires multiple steps, linked together manually, you make mistakes.

> Mistakes in communication result in lost revenue and excess costs.

Improving author experience – enabling the CMS to fulfill its purpose – provides a direct business benefit in communication quality and integrity. Whether you can free up individuals and resources is secondary.

In a world where reputation hangs constantly by a thread, subject to social-media mockery, poor content management can so seriously damage an organization's reputation as to put it out of business.

### *Professional tools for professional communicators*

Content management is about communication. And communication is the lifeblood of any business. Communication sells our products and services and supports our customers. Communication is arguably the most important function in any business.

Content management is a high-impact service provision within any organization. So, it makes sense that the people carrying out this function must be professional communicators. And professionals, whatever their expertise, need tools suited to their trade.

## Purchasing a platform

Content management tools have their own models and paradigms, which are not particularly suited to the mindset of the people who must use them. (See the section titled "The technologists' paradigms" for details.) This is in direct conflict with the purpose of a CMS, which is to facilitate a human process.

### *Technology purchasing process*

In the old days, any type of content management was considered to be an IT responsibility. After all, the system revolved around a big, scary computer – definitely part of the IT world.

And the IT department made its own rules, in effect dictating what the rest of the business could or could not do. If you wanted a system, you informed the keepers of all things technical, and they took ownership. Unfortunately, the best-of-breed solution they gave you met criteria that were not necessarily based on the functionality you needed. The IT department was buying a new toy that suited it's rules without asking what the new toy should do – not that the business really understood what it wanted anyway.[1] Most often, this disconnect happened because authors were not consulted on their needs, and if they were consulted, there was no content strategy to speak of.

As technology has become ubiquitous, more integrated into everyday life, other departments have gotten in on the act, making purchasing decisions for themselves and moving IT to a supporting role. This has, to some extent, been abetted by information systems becoming available as cloud-based services.

But one feature from the old model remains. Even when the right departments are buying, they often do not understand the full scope of the functionality they need before making the selection; they buy based on a checklist of features, too often based on the latest social buzz. They select the platform before anyone considers what it needs to achieve. Then, the platform's capabilities and limitations drive the customization. The business is forced to change its practices and processes to comply with the technology, which cannot be good.

> **We chose it because…**
>
> An organization I know is being forced by regulators to sell off part of its business. As part of the sale, they are required to set up systems for the spin-off business, including the marketing web site. They have already selected the platform. When I asked them about the selection, they admitted that they had not analyzed their communication requirements, nor had they considered the actual business needs or the platform's compatibility with them.
>
> As far as I can tell, the two most likely reasons they selected this platform are unrelated to their requirements or business needs:
> - They are already using this platform for another system.
> - They have spare licenses because a subsidiary just cancelled a build on said platform.
>
> I know the selected platform painfully well. I know what it cannot do. I know some of the basic requirements this organization will have for communicating their offers to the market. And I know the spin-off business will need to replace this system almost immediately.

---

[1] Am I exaggerating here? For some organizations, yes. But not for all. I have encountered many instances of IT-driven system selection that were unrelated to user needs.

## I need a new CMS

The reality of most content management implementations is that the process does not support creating an optimal author experience. The common sequence is the following:

- Someone identifies that a new or replacement system is required.
- Because the system requires a major capital expenditure, the IT department is tasked with fixing the problem.
- The IT department creates a list of functionality and features based on their perception of policy and end-user needs.
- An RFP is issued, and a new platform is selected.
- Designers and information architects are brought in to figure out how the system will work; maybe content strategists are brought in to determine what the content should be.
- The CMS vendor's implementation partner works to deliver on the promises represented by the RFP, while pushing back on the design and functionality.
- The new system is delivered to great fanfare.
- Within days – weeks at the outside – those responsible for maintaining the content wonder why everyone went to all that trouble. The new system is no better than the old platform; it does not help.

And so the process can begins again. This brings to mind Einstein's[2] classic definition:

> "Insanity: doing the same thing over and over again and expecting different results."
>
> —Albert Einstein

This does not need to be the case. There is another approach.

## Rethinking the process

If we add two elements to this process, the results become significantly different. (While either one alone would make a marked difference to the outcome, both are needed to get the best results.)

First, we must define the communications the system will carry, starting with the fundamentals: Why do we need the system? What is the purpose of the information exchange? From this, through a few more steps, we can identify and model the content required, at least in the abstract.

---

[2] Allegedly – variations of this quote have been attributed to others too. So if you prefer it being an ancient Chinese proverb, go with that variant.

Second, once we know what content needs to be managed, we can determine the content management paradigms[3] best suited to the authors – from initial creation by subject matter experts through copy-writing and editing to distribution through the myriad mechanisms, channels, and devices from which it is consumed. Who must interact with the content? How do those interactions differ at various stages within the process? What are the workflows associated with creating, editing, revising, and archiving the content?

With this additional knowledge, the outcome is very different. Instead of system vendors pitching a list of features and functionality based on matters that have little (ok, nothing) to do with what the system is intended to do, we have RFP criteria based on ensuring the following:

- The content managed through the system fulfills a purpose.
- The system enables authors to manage that content properly.

## *I need a new CMS, part 2*

The sequence changes to this:

- Someone identifies that a new or replacement system is required.
- Content strategists are called in to understand the communications purpose of the system and to model the content it must handle.
- Author experience consultants map the modeled content to organizational workflows in consultation with all affected parties, ensuring that the paradigms are appropriate to the audience.
- The IT department takes the requirements and determines whether a new system is required or whether the existing one can be extended to meet the content and authoring needs.
- An RFP is issued, and a platform is selected that is capable of managing meaningful communication.
- Designers figure out how to display the content.
- Information architects work with the vendor to map the modeled content and workflows into the system's paradigms.
- The CMS vendor's implementation partner delivers the content management, authoring, and workflow functionality they agreed to in response to the RFP.
- The new system is delivered to a user base of subject matter experts, writers, and editors who have been looking forward to a solution based on their needs.

---

[3] By paradigms, I mean the authors' mental models of the content they manage and the processes associated with those models.

By defining the author experience before choosing the platform, we accomplish two goals. First, the delivered platform is fitter for purpose and, thus, provides a better return. Second, the process benefits the content management community at large because CMS vendors learn that for one more client, they cannot use sales tactics that ignore the actual purpose of the system.

## Weighing author experience against user experience

We are all familiar with the concept of user experience: the need to focus intently on – even be obsessed with – the end user. Content strategist Eric Reiss defines user experience as "the sum of a series of interactions" between people, devices, and events.[4] However, in practice, there is a tendency to consider just the end user – usually a customer, an external party – as the user of record.[5]

With most CMS implementations, when author experience is raised as a concern, it is pushed back by the demands of end-user UX. The end users are most important; we must ensure that the system engages them, keeping them busy with our content (buying our stuff or racking up hits that work toward our advertising revenue).

How can the author be more than an afterthought when there are valuable customers to be wooed?

The mindset driving most UX is that unengaging content does not retain users, and a difficult-to-use environment encourages them to go elsewhere to fulfill their needs. If they have no choice but to use our poorly implemented system, they will become frustrated, leading to a combination of sloppy use and negative reputation.

That is a compelling argument to make UX your number one priority.

However, this argument ignores a fundamental detail. Authors are users too; the same psychological consequences apply to them. And given that the authors are responsible for managing the content that is supposed to engage end users, we cannot afford to use a platform that frustrates authors and keeps them from developing high-quality content.

---

[4] see http://www.fatdux.com/blog/2009/01/10/a-definition-of-user-experience/

[5] If this were not the case, authors would already be considered important users, and this book would not be needed.

The CMS must be valuable to the people who use it to manage content. After all, a tool that is valued by a worker is used by that worker.[6]

If you don't invest in author experience, the value of the CMS is severely degraded and the end-user experience is jeopardized. If you are spending money on a content management solution, it pays to consider which parts of the system provide the greatest overall returns. These will likely not be the quickest wins.

## Taking the lead in improving AX

As you can see, author experience is a critical element in managing content, and it becomes increasingly important as organizations embrace multi-channel communications. But the available tools and people's mindsets can get in the way. Chapter 4, *The Challenges to Good Author Experience*, contains examples of issues that hinder us – problems we must address to bring more tangible returns from the managed content.

But first, we have to deal with the question of responsibility. Whose role is it to ensure that author experience is given its due? Who fights for the budget? Who holds developers, implementers, and system providers to account? Who champions the communications process?

There is no simple answer, except to say: you do.

From a business perspective, anyone who understands the value of reducing risks associated with managing communications must be on board. If you own the budget for marketing, content management, or support, then author experience must be a high priority. When someone tries to siphon off funds to a pet project that will allegedly provide a quick win, the real price is in the stability and coherence of your content, putting at risk the very thing you are responsible for.

From a content manager perspective, I think it goes without saying: a good author experience makes your life simpler.

From a developer or implementer perspective, you have a responsibility to your clients, not only to provide them with what they explicitly request, but to understand the purpose served by the system you're developing. Don't trade authoring functionality necessary to that purpose for something else, even if that something else is the latest "in" thing that everyone else has.

---

[6] Paraphrased from Krista Donaldson: "A product that is valued by a customer is used by a customer." TEDWomen 2013, http://on.ted.com/ta4W

# CHAPTER 4
# *The Challenges to Good Author Experience*

Based on the definition of author experience, it should be a no-brainer: make systems work in a way that makes sense to the people who use them – particularly those who create and manage content.

If only life were so simple.

A variety of circumstances pose a challenge to creating a meaningful author-experience implementation. Looking beyond basic resource and budget issues, these challenges fall into four categories:

People challenges

- Knowing how to communicate
- The communication goal

Technical challenges

- The technologists' paradigms
- Content coupling

Process challenges

- Workflow that… doesn't
- Content ownership and editorship

Conceptual challenges

- Mental models
- Metadata

## Knowing how to communicate

It may sound odd that I identify knowing how to communicate as an impediment to a good author experience. The problem is the perception of competence and of value.

### *I can do that*

A significant proportion of the population are decent communicators. They can get a message across. They can deal with others, reading the micro-reactions displayed by the people they interact with. They are decent people-people. But can they create readable information?

The skills of the storyteller – the writer or journalist – are different from those used daily by most people. Writers employ structure in the form of headings and sections. They follow rules and conventions, working within limits. And when we read a professional writer's output, the meanings and hierarchies of the presented information are clear, even when we are not aware of the rules.

Amateur authors[1] do not understand the abstracted semantics of written communication. They demand WYSIWYG editors so that they can style everything in the way they feel is most appropriate for that particular message, without understand the relationship between the representation of semantics and the underlying meaning.

## *The value-add*

Just as amateurs think they know enough to communicate effectively, managers often fail to recognize the value professional communicators bring to content management.

As we have already established, content is just structured information, and the point of information is communication. Since communication plays such a pivotal role in business – be it attracting customers, maintaining relationships, soothing hurt feelings, or negotiating with suppliers – it goes without saying that professionals must carry out these functions.

Yet all too often, content management is an afterthought, something that temps or interns can do, because the good people are needed for more important things. And with authors so far down the food chain, no one bothers to give them decent tools to do their jobs, thereby compounding the issue.

When you hire carpenters or plumbers, do you expect them to use any old tools just because they earn less than you? Or do you expect them to have the most appropriate tools for the job, so that the results are of superior quality?

Likewise, giving those who manage content the tools they need to manage it – to move and manipulate, rearrange and compare, reuse, version, and reference – enables them to better craft the desired communication. Providing only a keyboard, a screen, and a single input field is like giving your carpenter a broken hammer.

---

[1] I use the term amateur not with respect to whether the individuals undertake these activities as part of their employment, but from the perspective of training and professional skills.

> **We've got the temps doing that**
>
> One client, a few years back, was re-platforming its intranet. Cleaning out redundant content was in scope, as was making retained content more meaningful. The "content team" they brought in consisted mostly of the cheapest contractors available: students.
>
> As a result, the process was slow. The lack of subject matter expertise meant that the content team could not make meaningful judgement calls about the content. They could not clean up content without context.
>
> After the entire team was let go – due to other project issues – another team was brought in at the last minute to copy content to the new platform. Any attempt to clean out or re-write the content was abandoned.

*Solving the knowing-how-to-communicate challenge*
The incorrect perception of the value of competency in managing content can be resolved only through education. There is no technical solution.

## The communication goal

The second challenge is somewhat abstract. To understand this challenge, we need a little background.

*In the beginning...*
As humans, we communicate in three basis ways:
- Acting[2]
- Speaking
- Writing

Writing started with simple imagery. The canvas used was found in nature: the ground or a cave wall. As writing evolved, so did the choice of what was written on; writing became transcribable.[3] From these clay tablets, wood etchings, and basic scrolls, perpetual information storage evolved to mass-printing.

---

[2] Yes, most acting now also involves speaking. But in its most basic form, acting refers to imitative representation. It is visual.

[3] Transcribable writing is writing on a portable medium, one that allows a copy to be created, moved, and copied again. It can be checked for errors – a process that would be unreasonably cumbersome between cave walls.

Through the evolution of written communication, one thing has remained constant: the consumable text is static; it cannot adjust to the audience. This differentiates it from the other two forms, which allow for real-time adaptation. In a spoken recounting, the bard can embellish the tale to take into account the context and the audience's reaction. Some forms of acting, such as pantomime and comedy, involve audience participation.

## Content in a digitally distributed world

As we entered the computer age, the static nature of the written message persisted. The sent message was the received message, packaged one way and intended to be consumed in one format. This mindset, derived from print, persisted because the presentation – as advertisers were keen to remind us – was as much a part of the message as the content itself. Indeed, the medium is a form of content.

There were attempts in the early years of this millennium to provide contextually dependent delivery. Most of the time, however, these delivery methods were restricted to choosing what content to deliver, rather than choosing the structure or form of the content. Some undertakings were more ambitious and successful, but they remain locked up in the enterprises that invested so much in developing them or were restricted to specialist industries. Even now, contextual delivery is still in its infancy.

It took the emergence of the high-powered, touch-enabled phone, with a screen large enough that it takes several seconds to read its content, for the first real rumblings of a shift to occur. The forms of display suited to the large screen of a desktop computer or laptop – effectively an electronic version of static-layout paper – failed to translate to the physical constraints of the new screen.

Consequently, the concept of serving variants of the same message, with differing emphasis or calls-to-action based on information about the audience, has gained traction.

## The goal

Everybody wants his or her message to have the reactive characteristics of verbal conversation; that is, to adapt to subtle interactions and to be personal.

However, this goal is at odds with technology and human nature.

On one hand, those responsible for the message refuse to trust a computer to do anything more than serve up a pre-approved message from a selection of possibilities. On the other hand, even though audience-segmentation criteria enable the definition of dozens, if not hundreds, of personas, we cannot devise, approve, or manage enough message variants to address all of them.

> **The more-content dilemma**
>
> One project I was involved in a few years ago for a pharmaceutical company called for the creation of many small snippets of text aimed at encouraging and rewarding the user's behavior. In itself, this was a good idea.
>
> However, because of the industry, every piece of text had to be approved by the regulatory department, and we had to provide source attribution. The list of approved sources was minimal: a few pamphlets and articles. As such, we could not generate enough content to cater to the diversity of possible, relevant contexts.

In theory, we could create a series of smaller content bits that the system could amalgamate using *multi-axis profiling*. However, if those responsible for communication insist on approving every possible variation, this option goes nowhere fast.

The demand for personalized communication competes with our limited ability to create a manageable set of message variants.

## *Solving the communication goal challenge*

Two approaches address the challenge of aligning what people expect from digital communication channels and what those channels can deliver. One sets realistic expectations and communicates them clearly. The other waits until systems have been developed that can meet expectations. I recommend the first approach; the second is some years off.

# The technologists' paradigms

Our third challenge derives from the way most content management systems are built. Some systems are built to fulfill a specific need: they start with business requirements, content models, and authoring paradigms and only then are they implemented. These are effectively single-use systems, tightly aligned with the initial need. They are fit for purpose. They do one job well, but cannot be easily adapted to other roles. They rarely escape the environment they were developed for.

The majority of systems, however, are built as productized technology: generic designs for storing or manipulating content, neither tied to a specific task nor appropriate for the needs of any content.

As important as technology is, it must not be the driving force behind content management. Used correctly, technology facilitates, rather than drives, the process.

## *The CMSes dirty underwear*

When a technical idea evolves into a content management solution, it is usually the result of an issue with the storage and use of information. The programmers are looking to solve a technical problem such as performance, stability, or capabilities. The internal paradigms for any CMS will differ depending on the issues its developers were trying to solve.

The problem occurs when the internal paradigms – for example internal content storage models – are surfaced in the management interfaces and dictate the functionality and usability of the author environment.

For example, if you were running a targeted campaign, would you want to use a system that required you to manage audience profiles in one place and promotional content in a second place, to use placeholders in your main content to indicate where the campaign variants are displayed, and to link all the parts together in a fourth place[4] – all of this without the capability of automatic linking between the parts? Or would you expect a seamless experience?

These interface problems are not just problems for authors. Sometimes, developers find themselves in the same boat.[5]

Of course, the developers' content storage, structuring, and management paradigms make perfect sense. They are the core of the implementation logic; they improve performance, stability, or functionality. And because the developers understand the model's logic, using that conceptual approach to manage content seems quite reasonable to them.

Unfortunately, these developer-designed models rarely make sense to those who use the system to manage real content.

---

[4] You may think I am making this up as an absurd example. Unfortunately, I'm not. This really is (or at least it was a version ago) the workings of a popular, and expensive, system.

[5] Garann Means expressed this beautifully in a Pastry Box Project post in June 2014 (https://the-pastry-box-project.net/garann-means/2014-june-16).

**What type of content is that?**

One system I have worked with manages content in folders and pages. Everything is a page. Within pages, it supports components: child elements.

If I want to create a list of options and make them available elsewhere in the system, I have four choices:

- Create a folder and populate it with a series of pages, one for each option.
- Create a container page and add instances of an option component to it.
- Create a page with a special, embedded multi-input component that allows me to add rows, one per option.
- Borrow the tag-management interface, which doesn't look anything like the page management interface, but creates future dependencies.

None of these is particularly clean, especially because I am managing a list of options. While the system may treat the list as a page or as tags (and tags are really just pages with a different interface to manage them), it does not make sense for me to have to look in the content "pages" to edit this list.

## *Keeping track of associations*

The problems can get worse when it comes to content association paradigms. What is your thought process when you want to identify the creator/owner of an article? And how must you twist that thinking to accommodate your content management system's data model?

Figure 4.1 – Differentiators for identifying a person

You want to identify John Smith. You know John as a person. If three people in your organization have that name, you apply a cognitive differentiator : appearance, role, or something similar.

The system expects you to identify the right John Smith based on the path `/content/site/people/s/jonathan.p.smith.xml`.

# 28  The Challenges to Good Author Experience

You probably didn't know that John's real name is Jonathan, because he never uses that spelling. And the less said about that middle initial...

Associating a person with a path makes perfect sense in the context of storing and processing the information. But surfacing that paradigm to authors using the raw path does not.

## Whose paradigms matter?

Technologists and developers understand the models and paradigms used within their system. They can (and, given the opportunity, do) explain in excruciating detail how the model works. But they often have trouble understanding the difficulties others experience adapting to these paradigms. They struggle to see why their models are unintuitive.

So, must authors adapt to the internal content models or can there be a layer that translates between business thinking and storage paradigms? I think that the purpose of the CMS answers this question immediately: it exists to facilitate the process, not to enforce its information model on the authors.[6] Also, enabling authors to manage content in a way that makes sense to them reduces content risks.

There is a third interested party: the budget owner. Sometimes, the cost of reducing content risk is hard, or impossible, to justify. The problem often is that, by themselves, individual author experience improvements may not provide sufficient value for their cost. However, discrepancies between author paradigms and system content models are multiplicative in their impact; each one makes all the others worse, too. So each improvement has a bigger impact than you might assume.

Keep in mind that technologists and developers have had many years' experience explaining how expensive it would be to adapt their interfaces so they make sense to authors. You need to challenge them on this.

## The upgrade path

One of the most cumbersome issues you face when customizing a CMS to create a better author environment is managing upgrade paths. Most commercial CMS vendors make the client responsible for the cost of updating a customization when the underlying platform is upgraded. If

---

[6] In 2005, Lynda Gratton and Sumantra Ghoshal from the London Business School published research showing that simply adopting best practices is not best practice. High performance organizations embrace unique signature processes: they do things in ways that make sense to them, instead of blindly adopting others' models. See their article at: http://sloanreview.mit.edu/article/beyond-best-practice/

the next version of their CMS no longer interfaces cleanly with your customized authoring environment, it is up to you to make the changes. They are not responsible.

> **Don't give the client the right solution**
>
> I was writing a system customization specification for a client a couple of years ago. They had told me what they wanted and how they needed it to work so it made sense to them.
>
> When I showed a draft to the system integrator's senior partner overseeing the project, I was told "No. Don't suggest that." The reason was simple: the client's needs required an approach that did not align with the system's paradigms. The partner was confident that making the solution fit the client's needs would guarantee they could not upgrade to the next version of the platform without completely re-engineering the implementation.
>
> As a result, the client had to live with a more cumbersome interface.

Upgrades to the CMS may include performance enhancements, security improvements, and new features that better align the system with standards. Clearly you don't want to be prevented from deploying an upgrade because your implementation is too customized.

This means that any customization could incur future re-implementation costs – an uncertainty we could all do without.

The threat that paradigms may shift with the next version, requiring you to abandon or update your customizations, holds you hostage. Even though you know your needs, you don't know the value of your current investment or how much it might cost to upgrade when the next version of the CMS arrives.

## *The flexible system*

Faced with the discrepancy between author mindsets and developer paradigms and threatened with the possibility that customizations may be made obsolete by the next upgrade, you could easily despair and decide that it is simply not worth the effort. But we are talking about managing content, communicating. We have no choice; we must find a balance.

We need a CMS that provides the flexibility to manipulate content using paradigms that differ from the storage model. We need an interface that enables us to build a translation layer between the storage model and the authoring interfaces that is not affected by changes to the underlying system. Unfortunately, creating such a complex interface would probably require sacrificing some of the performance benefits the technologists' models aimed to achieve. Developing such a system, without sacrificing performance, would be a mammoth undertaking – one that no one has yet seen fit to invest in, despite the obvious benefits.

And before you start thinking that such an interface would allow you to swap out the underlying CMS while retaining your optimized authoring interface, remember that each vendor would have its own model for how the translation layer interface would work.

### Solving the technologists' paradigms challenge

Overcoming the conceptual misalignment between how content managers think about their information and the paradigms used in content management systems is at the core of author experience. Practically, this calls for authoring interfaces that are mapped to author needs and expectations. Some aspects of this are discussed in the section titled "Fit-for-purpose language" and the section titled "Content accessibility" in Chapter 5.

## Content coupling

The fourth challenge is one of the most commonly cited problems with current CMSes: the degree to which content, presentation, context, and behavior are tied together. Per-device differentiation makes things even more convoluted.

For clarity, let me quickly define each of these:

- **Content:** any text, images, video, decoration, or user-consumable elements that contribute to comprehension.[7] Basically, the elements of the communicated message.
- **Presentation:** the spatial and temporal arrangement of elements of content and any wrapper elements required by the delivery medium.
- **Context:** the specific situation for and to which content is being delivered. This relates to many details, including device, channel, user needs, behavior, previous interactions, and even state of mind.
- **Behavior:** the mechanics of interactions; how a change in the presented content will affect a user's actions.

Different CMSes suffer from these issues to different degrees.

---

[7] Definition from *The Language of Content Strategy*, XML Press, 2014.

## *Single output (presentation coupling)*

Many web CMSes were designed to manage content for a single output, a legacy that hobbles them in the modern era. In such cases, it is all too easy to mirror the presentation technology's paradigms in the content storage model, which then carries through to the authoring environment. The most common manifestation of presentation coupling is *rich text* editing, where presentation instructions are embedded in the content.

One consequence of presentation coupling is a lack of structure, which means everyone gets the exact same content, in the same-sized chunks, laid out in the same way, in the same sequence. A second consequence is that redesigning output requires manual changes to every piece of content; the content migration portion of a web redesign can take so long that it is still in process when the next redesign begins.

While the rise of mobile devices is gradually changing this, the situation persists.

> **Lessons unlearned**
>
> As part of a re-platforming project for a client, we found that their existing content was mostly on one no-longer-supported technology. Within that content were several presentation styles, some left over from previous redesigns and some from departments doing their own thing.
>
> One of the new system's next-phase requirements was mobile optimization.
>
> Despite the obvious lessons in presentation-coupled content, they could not understand the value of managing their content stripped of presentation. They chose not to invest in getting the structure in place for the future.
>
> As a result, the mobile optimization project will end up being bigger than the original project because everything must be re-implemented. If, that is, anyone will sponsor such a massive undertaking.

## In-line rules (context coupling)

The coupling of content with context is prevalent in current technical communications platforms, especially in *DITA*, a standard that many vendors are adopting. Among other things, DITA lets you apply contextual rules to inline content – enabling, for example, a unique version of a sentence for each audience.[8]

By exposing content variations using in-line mark-up, these interfaces severely limit the contexts that can be reasonably managed (i.e. understood and visualized) by an author.[9]

Unless the authoring environment includes a context filter, this model can overload the author, displaying all options for all audiences, stacked together and, as likely as not, in no particular order. Additionally, the flow of text is polluted with code to indicate that particular sections are to be displayed only to audiences that meet particular criteria.

Given that this mark-up could run to almost half a line per instance, how is one supposed to understand a paragraph that has almost as much contextual mark-up as content? Once again we see the surfacing of technical paradigms to authors.

## Treated like sheep (behavior coupling)

Because their architectures lean towards presentation or context coupling, many CMSes are built in a way that assumes all users will share the same behavioral interactions with the content.

The way users read a long-form article is a good example. In a desktop web or print environment, a reader probably expects content laid out as a single sequence, with the media in line. But in a mobile environment, the reader might want the media elements to be accessed only on request, using a hub-and-spoke interaction.

A CMS that assumes one style or the other and forces a one-size-fits-all behavior model will end up in an untenable position, especially as new platforms and media become available.

---

[8] This feature of DITA may not be considered particularly interesting by the community that embraces the technology, which explains why I call it out here; it is not exciting enough to receive attention and be improved.

[9] Noz Urbina (@nozurbina on Twitter) of Urbina Consulting has written a pair of articles on this subject: http://dclab.com/resources/articles/conditional-content-top-tips-part-one

## One device at a time

One of the "solutions" developed for the content-coupling problem is to separate content management by output device. The diversity of output devices currently available should be a good indication that this is not a viable approach.

The growth in device and platform availability is forcing CMS providers to look at structures that do a better job of decoupling content from its presentation, context, and behavior. But the process could be sped up if we pushed vendors harder.

## Solving the content coupling challenge

Solving the problems related to content coupling requires a two-part approach. One is technical, as covered in the section titled "Associative structured content" and the section titled "Rules-based presentation" in Chapter 5. The second is training authors to think about content as reusable elements, separate from presentation, context, and behavior.

# Workflow that... doesn't

The fifth challenge with content management is workflow.[10]

## Stop, go, get confused

The most common perception of workflow within content management is a structure to ensure content governance – a series of hoops content creators must jump through to publish their content.[11] This usually takes the form of a review process, with stop/go logic at each step. It is this aspect of workflow that most CMS vendors concentrate on when extolling the power and flexibility of their workflow systems.

In its basic form, such an approval system is straightforward (see Figure 4.2): The author creates something, submits it for approval, and it is either approved or rejected by the approver. If approved, it is published. If not, it returns to the author for correction.

---

[10] How anyone can have the audacity to call them workflows, when they so clearly do not work, nor do they flow, I cannot figure out. (Reviewer's quip: "it's called work not because it *does* work, but because it *is* work.")

[11] This aspect of workflow, and ways to work around it, was written up nicely by John Eckman of The CMS Myth: http://cmsmyth.com/2013/09/the-cms-workflow-myth/

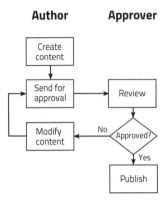

Figure 4.2 – Basic workflow

Adding one more hurdle – a second approver – makes the process much more complicated (see Figure 4.3).

The process for the first approver process remains as-is. But if the second approver rejects the content and it goes back to the author to fix, who does it go to for re-approval? Does the re-review process depend on the scale of the changes? The first reviewer signed it off in its original form, and the second probably does not want to micromanage the publication process. But if the changes requested are significant, the first reviewer may want to consider it as a new piece. The process is messier still if the first approver has an editorial role.

Are you confused?

Imagine the headache resulting from authors, supervising editors, and three rounds of approval, where each level consists of teams rather than individuals. Eventually, no one has a clue what is going on, what has or has not been published, or who is responsible for the holdup in publication (oh, sorry, he's on leave).

# The Challenges to Good Author Experience 35

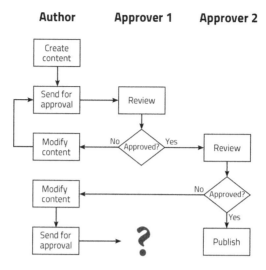

Figure 4.3 – Two-reviewer workflow

> **Approvals-coupled functionality**
>
> Here is an even more excruciating case of workflow getting in the way. I was working on a platform for a pharmaceutical company. Everything they did needed to go through several layers of approval, including legal and regulatory. This applied to both content and functionality.
>
> Because of the platform, the approvals process created issues. The database that contained the functional code and associated data did not support partial transfers from the author environment to the publish environment. The entire database had to be migrated in one operation. But by the time a piece of functionality had gone through the whole approval process and was signed-off to go live, there were two months' worth of new functionality in the queue. The code and data for the newly approved functionality could not be put into the publish environment because the database also contained code and data for unapproved functionality.
>
> This "workflow" process, combined with the technology, effectively guaranteed that no work could flow, which was the case for at least the four months that I was working on the project after initial release.

## The approvals escape hatch

Except in the most basic form, any approval system coded into a content system is bound to run into problems. All these layers of approval, revision states, annotations, branches, forks and notifications, and optional paths become dependent on abstract concepts like the scale of the change.

The larger the process, the slower it becomes. But business needs do not tolerate delays. Urgent messages must be published immediately, not next month. Approval cannot be held up by someone who is in a meeting, let alone gone home for the day.

And despite everyone's best intentions, even with a dozen layers of review (or maybe because of them), critical errors will slip through. Information will be published before an intern spots a typo that needs to be fixed without delay.

As such, work-arounds are required. And, of course, the existence of work-arounds corrupts the official process. Soon, everyone uses the work-around as the default process because it is faster.

## How work flows

Approval processes are not what content workflow is really about. Treating workflow like an approval process just formalizes in software a process that no one had a good grip on before it moved into that realm. (There is no silicone bullet.)

Real workflow is about how work cascades based on the relationships and dependencies between pieces of content. Altering one piece of content affects another. And when content expires, other content can be left unused, orphaned, or in need of retirement. Information is complex: it contains associations that are dynamic and multiple.

> **Copy, paste, edit, find, change, repeat**
>
> I have several times been involved in situations where subject-matter experts created content in Word. This content then got added to the content management system. When the SMEs changed their minds, they sent a revised Word document (only sometimes including tracked changes).
>
> Someone was then expected to find the differences between the two versions and apply those changes to the system.
>
> Is such a choice of technology really suitable? Who does it enable?

A change to one element of information affects other elements. Dependencies must be followed. Most content management systems handle dependencies only at publication time. When one element requires another element to be published, the system ensures that both elements get published. However, most content management systems can't remove obsolete and no-longer-referenced elements. Nor can they maintain the associations that show what information needs to be updated as a consequence of a source information change.

Real workflow is about process and movement. It is a multistage sequence that starts when you determine that information needs to be conveyed and continues until that information is retired. Content is written, checked, edited for the audience, re-edited for context, and broken up so it can be rebuilt for the delivery channel. Workflow ties these steps together so that when the core information changes, updates propagate – with or without human intervention.

## *Solving the workflow challenge*

Resolving workflow-related issues requires changes on two fronts. First, we must change our expectations of what workflow means and how much of the process can be encoded in the system. Second, we need technical enhancements to deal with the real processes by which work flows. These technical enhancements are discussed in the section titled "Content management tools" and the section titled "Self-aware content" in Chapter 5.

## Content ownership

Our sixth challenge relates to content responsibility: who owns, manages, and maintains existing information. While this could be considered as a part of workflow – and indeed, it profoundly affects workflow – it is large enough to warrant a section of its own.

## *Who owns the content?*

The question of content ownership is important to ask, but not easy to answer. While it is usually easy to determine ownership in small organizations, the typical single point of ownership model does not scale. We need to understand which aspect of the information we are dealing with to assign ownership. The structure of the message? The words and media presented to the consumer of our communication? The source material?

> **Owning is not enough**
>
> Even for small organizations, the single-point-of-ownership model is fraught with dangers. One client rebuilt its web presence about ten years ago. It added great content, relevant to its activities. Now, a decade later, it is replacing the site. And the content that was created back when the website was launched is still there, in its out-of-date form.
>
> Another client, in re-platforming its system, decided it needed all content to be reviewed on, at least, an annual basis. It wanted owners to be notified when their content was due for review and if it was late. While a laudable concept, their approximately 10,000 pages generate around forty expiration notifications per day! (It is even worse if we consider that everything was prepared in a short window, so for several years, most notifications will be clustered in a three-month window.)
>
> It is not enough to assign ownership. The people responsible for content must have the time to do their jobs.

The ownership aspect of governance is rarely identified in a meaningful way. This is a business-process issue rather than a technology problem. It is a consequence of information in the pre-digital economy being not nearly as reusable as information can be today.

What does ownership mean? What does it imply responsibility for? The initial idea? Content generation? Worth determination and measurement? Usage? Review? Archiving or retirement?

And are there multiple owners with different levels of responsibility?

## *The evolution of ownership*

A core aspect of information ownership, I believe, is the evolution of responsibility as information is transformed. This is the ownership aspect with the most significant impact on workflow.

Ownership evolution is critical to information management. Whether it is managed as an integrated function within your CMS or handled outside the system is another matter. Fundamentally, ownership evolution is about bidirectional traceability between an idea, the source (subject matter expert) material, and the various drafts, iterations, and manipulations of that content through all publishable versions.

Imagine a product being sold. There are technical specifications and designs. There are function and feature lists. There is imagery. There are marketing themes. There is the sales material itself. And there is support information. If all of these are maintained in a way that links them together, a change to the underlying specifications can propagate – either automatically or by informing the relevant parties of the need to update – all the way through to the sales and support material.

It is hard enough for most people to imagine these dependencies with only one type of content. With many types of content, additional dependencies come into play that make for a complex environment. It is messier still when information is managed separately across channels.

This association of idea with design with features with publishable material rarely exists because, too often, content management systems are seen as repositories for published material rather than systems that deal with content through its entire lifecycle,

## *Owning the archive*

While on the subject of ownership, let us also consider the ownership of archived material. By archived, I mean content that is still available but not considered current. The classic example is a news organization whose back issues are still available for reference but are obviously no longer today's news. While much of this material will fade into obscurity, it never becomes completely redundant. A story from a decade ago could rear its head once more when the subject is again newsworthy.

The idea of archive ownership is moot here. Considering the resources required to keep track of everything, the rate at which people change careers, and the monotony of overseeing a growing but rarely referenced store, the single point of responsibility model is not viable.[12]

The best ownership is dynamic, by referenced association. If a new article references old news, then any update to the old to reflect the new changes becomes part of the new article's workflow.

---

[12] In 2014, the New York Times published a correction to an article from January 1853 (http://nytimes.com/2014/03/04/pageoneplus/corrections-march-4-2014.html). The correction resulted from current events that led a reader to spot the error. Archive ownership is clearly impractical at such a scale. Responsibility, as the New York Times demonstrated, belongs to those dealing with current issues. Archived material, without a dedicated owner, is maintained only when a current issue relates to it.

## Syndicated content

As part of the ownership discussion, we must also consider how our information spreads. In a digitally distributed world, our information can – and does – go everywhere. We cannot control this. When we create an element of information, it will be aggregated, whether we like it or not, into third-party information streams.

Regardless of whether we view this as theft of intellectual property or as free promotion of our brand, we want to control our information's lifecycle. If syndicated information becomes obsolete, especially if we have an updated version we believe people must have access to, we want to present only current content.

And here's the zinger: we don't care if we're being ripped off or promoted; we don't want old versions of our information in the wild.

## Solving the content ownership challenge

The issues relating to responsibility for content can be dealt with mostly through technical implementations designed to make content more manageable. Some of these are discussed in the section titled "Content management tools" in Chapter 5.

# Mental models

Number seven on our list of challenges is predominantly a human issue, but it leads to technology being unable to deliver a workable solution. Many technical approaches map the simplest mental models people have about their communications, thereby reinforcing those models. But those models do not serve the dynamic nature of distributed content.

## The page

The worst mental contamination comes from one word: page. In content management, almost everything is a page. And when we think page, we see a sheet of paper with text on it; the information is static.

The issue is the word itself; the word page evokes a concept that directs how we think about the information it contains.

It does not matter whether the technical reality is a well-structured content type, designed for reuse with a range of rendering methods. As soon as we use the term *page*, we create the perception that we are talking about a sub-element within a sequence of sub-elements – something intended to be consumed after the previous page and before the next. It is one static snapshot within a series that, when replayed in order, creates the illusion of motion. It generates story through that sequence.

Of course, the word *page* is so ubiquitous that no one can provide an alternative. If a different word were to be the default term within a particular environment, people would still think of pages and refer to the structured entity as such. They would then complain that the system uses some obscure word that is just meant to confuse them.

The term now defines the reality, carrying with it the baggage of past meaning.

## *Selfishness of pages*

A significant part of the problem with the page concept is that, despite a mental model that implies that pages form part of a narrative, they are too often seen as being in a ring binder, each an individual, self-contained entity. This leads to the idea that pages can be added and removed with impunity. They cannot; the integrity of the information network must be considered when the content set is modified.

> **Don't mess with one of us alone**
>
> I wrote the original draft for this book in Word. It was tempting to just put down every thought that came to mind, each idea a page. This would have resulted in a book that jumped from place to place, lacking coherence. It would have been unreadable by anyone but me.
>
> Instead, I imposed structure on the book: introductory content, definition of terms, outline of challenges, practical information, and a look at the future. I used further structure based on the chapter themes. Sections can't be shuffled or rewritten without considering their impact on the surrounding material.
>
> And in Chapter 7, having written the original draft, I realized that a better structure – a concept, the reason, and the working approach – would get the ideas across better. I needed consistent patterns within those pages.

Also, the concept of a page as a pre-sized container is troublesome. Content that persists beyond its immediate form will benefit from structure and granularity even if it currently makes sense to manage that content in a page-centric manner. When the delivery model evolves, do the current page boundaries persist? And what about the consequences of presentation coupling?

An inherent problem with the page as a self-contained, standalone entity is the consequent perception that all users, of all types, over all interfaces, must receive information in the same size bites.[13] But a natural sequence in one medium may be cumbersome and disjointed in another.

## *Thinking about thinking*

Have you ever tried analyzing how you think about what you communicate? For most people, this is difficult. Nebulous feelings come into play, and the only aspect of the communication that can be reasonably analyzed is the final output: the single block of script, which equates to a page. Hence, thinking about communication as the whole page is the dominant model.[14]

The building blocks of knowledge we pull together to express narrative are not generally understood. To construct a sentence to express an idea, we consider – albeit without being consciously aware of the process – various words we could use, sentence structures, analogies, and similes. If we have trouble understanding the elements we aggregate to form a message, if we pluck the content from a cloud of vague knowledge, how are we supposed to create digital systems to manage the pieces? How are we supposed to codify a process that we are unable to describe?

We are attempting to communicate through written media (and yes, this includes video). We want to incorporate the dynamic attributes of speech or acting. But we don't understand how to codify the dynamism of human interaction. We don't understand how to break knowledge into fragments and recreate the logic that brings it together into coherent communication. We may no longer be limited by the single, defined order of older media and we may have some ability to contextualize for the audience, but for now, dynamic, computer-generated interaction exists only in the realm of science fiction. So our thinking is pushed back to the page.

## *Writing in sand*

The issue of page-based content is difficult to solve. The content community has generally embraced the mindset of the page as the base entity.

---

[13] This concept of one page in one environment mapping precisely to one page in another is called *page parity*.

[14] Yes, there are people – various scholars, for example – who can think about sub-elements of a block, about conditional variances within the structure of a single sentence based on multiple possible sentences that could surround it. But those people are specialized in their perception of communication. Most, including professional communicators, consider the value of the whole message, including its presentation.

It does not matter that many content practitioners rail against the idea of the static container; they too fall into the trap of using this terminology because it is the language their audience understands.

To resolve this, we need a model based on flow: a sequence of smaller grains that create a story. A river of grains that interrelate, though all may not be used in a particular telling.

So far, this could be seen simply as a cluster of miniature pages, except these snippets are too small to be called pages – they have no identities of their own, and they cannot be displayed as standalone content. They exist in an abstract mix where they surface only in the context of some larger sequence, reused outside their original manifestation.

> **Grains of content**
>
> Several projects I have been involved with over the last few years employed granular content; not on the scale I would have liked, but for small elements of interface copy (known as micro-copy). Each was a stand-alone entity, referenceable, and described.
>
> From a technical perspective, each of these elements could be considered as a page unto itself. But there was no case where such an element would be presented to a user out of some larger context; it would have been meaningless.

Moving to this granular content model will take time, will require that systems be developed based on it, and will stretch many authors' ability to conceptualize the information the systems contain.

## *Adaptive content*

As I have mentioned above, information contained in digital storage systems is destined not for a single output, but to flow into countless containers. Some of these are of our own devising: our platforms and devices. Others are beyond our control – beyond, even, our awareness of their repurposing.[15]

---

[15] Beyond our awareness of repurposing? This means that the information we manage is in a pure data form, made available for others to reuse in their own applications. We cannot know how it will be used, let alone presented. All we can rely on is the structural integrity of the information itself. Think, for example, of the data feeds provided by a major sporting event such as the Olympics. These feeds provide a live stream of participants, results, and even background data to hundreds of media outlets around the world.

The process of reformatting the same content for multiple deliveries is called *adaptive content delivery*. Content that is fit for adaptive delivery is called *adaptive content*.

Adaptive content embraces myriad forms simultaneously. It is not one or the other, it is all; it is fundamentally polymorphous. How the content functions and behaves is dictated by the environment through which it is experienced by the user.

But how is a person – someone who thinks in terms of the message presented in a single context – supposed to understand such content? How can we preview it? Which representation is true? With many known forms, and countless unknown ones, there is no way we can individually preview all variations and transformations of a block of information. And even if we could, how would we manage a situation where one instance does not work quite as we anticipated without affecting the majority that meet expectations?

How can we determine that our information is appropriately defined?

This is the challenge of structured input and rules-based presentation: we need people who can understand the transformations we intend to apply and who can see how the various presentation models will behave without needing to see every specific instance.

We all want our published information to flow and adapt, yet it flies in the face of human nature to let go of presentation control sufficiently to allow it to do so.

## *The content complexity issue*

If we model content as very small grains, with rules for how it should be presented to its audience, some properties would probably emerge from those rules and interactions that would be counter to the message we want to convey. This is the nature of complexity: many elements, all of the same type, following simple rules. But from the interactions, unexpected behaviors emerge on a larger scale.

So, using complexity to aggregate the message is not viable until we have modeled the human thought patterns that allow us to communicate dynamically. But maybe we can use the recursive model – itself an aspect of complexity[16] – from a structural perspective. At whatever scale we

---

[16] Do not confuse complexity with complicatedness. Complexity is something very simple, which repeats recursively within itself on a smaller scale. Complicatedness is just messy.

view content, whether the entire article, a section, a paragraph, or even a link, the structure is the same. An element of content is made up of three parts that combine to present it correctly: actual data (the words or media), a sequential pattern for presenting the data, and conditional rules to determine whether to include it. And each of these parts is just another instance of this same generic structure, repeated recursively until some fundamental primitive elements are reached; one simple model, repeated within itself.

> **The complexity-structured content model**
>
> More than a decade ago, I started documenting a new content management system. The fundamental design relied on two principles: first, that the structural definition of content is, itself, content that can be fully represented using the same model as the message content and, second, that the content, presentation structure, and contextual model are all "content" – interchangeable.
>
> While I have created a basic proof of concept – it works – my aversion to writing code means that it has not yet been implemented.
>
> I was expecting the content management industry to have caught up by now, but they are still years away. This means there is still an opportunity to leap to the front of the CMS marketplace with a system that is designed to be abstract. That is, a system that uses a recursive model to ensure the authoring interface is defined as content, appropriate to authors' needs and paradigms.

We can use the principles of complexity to structure stored information. Instead of trying to solve the problems of the dynamic conversation, let's make the storage as dynamically adaptable as we can. It will be needed when computers can finally hold dynamic conversations.

## *Solving the mental models challenge*

Mental-model issues are fundamentally about how people think, and the way to deal with these issues must be based on educating people. This said, having content management tools that are designed to deal with content in a structured way will certainly help, and the principles of self-aware content will likely make things easier.

## Metadata

The last issue I will deal with in this chapter is *metadata*. Metadata is information about content that is used by either the authors or the system to determine how content is presented or what content is presented. As we move away from the problems of content coupling, metadata is the glue that holds everything together. No other model can enable us to communicate effectively in a massively distributed information economy.

Metadata can be free form, like the *tags* on many websites and blogs, or it can be organized into *axes*. For example, a press release might have metadata axes for author, release date, and product, among others.

### Too much information

Despite the need for metadata in content management, this is an area that often goes wrong.

Sometimes, to categorize content in a meaningful way, you need to use many metadata axes. But if the axes are not well considered – especially if authors know the metadata is never used or do not understand how it applies – the axes go unpopulated or are filled with whatever meaningless values the author can think of to get past the pain point.

> **Axes vs tags**
>
> There is a general understanding, perpetuated via blogging and other social media, of content categorization by tags: add a set of descriptive words and use them to understand what the content is about. So why do I talk of multiple axes of categorization?
>
> Basic tagging uses the premise that all descriptive labels are equal. Multi-axis categorization gives semantic meaning to the tags. It can specify that a tag must relate to a particular concept; not just any tag will do.
>
> For example, I recently watched two movies: *All Is Lost* and *The Conspirator*. With simple tagging, both would be tagged "robert-redford," because Mr. Redford was a major player in both. But doing so would lose a key bit of information. For one film, the "robert-redford" tag belongs in the actor axis; for the other, it belongs in the director axis. These are not the same. Using only a single, generic axis would confuse.

In other cases, especially when it comes to descriptive metadata, the available structures and options may not provide sufficient scope to be meaningful. For example, when articles are written for a publication using a very tight theme, the metadata values given to each article can easily end up being the same, thereby eliminating any value that might have been present in the metadata.

> **Let me categorize that**
>
> I have been guilty of both mistakes: using filler metadata and publishing material that I did not – could not – sufficiently differentiate with metadata. The blogging platform I have been using expects (though does not demand) tags. Lack of tags means the content is hard to find, so in go whatever terms I can think of that describe the article. Which happen to be – for the most part – exactly the same terms as the previous article. And the one before that. The blog has a theme, and every meaningful term I can think of is relevant to almost all the articles.

## *Types of metadata*

If you have created or managed metadata, you know that there are several ways it can be specified.

We can select from a fixed list, have a pick-list that new values can be added to, or enter entirely free-form values. Some metadata fields expect a single value, while others accept multiple values, and those multi-value metadata fields may be ordered or unordered. Some metadata is author-generated, while other fields are populated with values that result from computations, algorithms, or interactions.

Most commonly, though, our first distinction comes down to the simple question of whether a field is optional or mandatory.

There is nothing particularly difficult about the requirements of mandatory metadata, except… Well, this is not true. There are countless digital interfaces where an input form will ask for something as simple as an address and demand a county, state, or province, even though many people live in places where these geographical distinctions have no meaning.[17]

---

[17] In the UK, it is common to be asked for a county as part of an address, despite the Royal Mail stating that this is redundant. London, a city of eight million, does not belong to a county; the nearest equivalence would be to call it "Greater London." I cannot be the only one to think an address including "London, Greater London" is absurd.

## Meta-metadata

Things become significantly more interesting when we put the meta in our metadata. What happens when the identification of a mandatory categorization pattern for an element of information within our system is the result of one or more other elements of metadata?

We can easily determine what to do when creating content of this type. First we must specify the data values that drive selection of the subsequent pattern. Only then can we identify what fields are mandatory in the rest of the definition. This adds a layer of complexity to the metadata, what I jokingly refer to as meta-metadata.

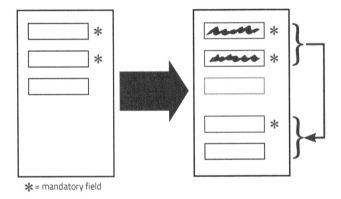

✳ = mandatory field

Figure 4.4 – Metadata structure dependent on supplied metadata

In Figure 4.4, when the two mandatory fields are filled, their values cause the third field to be disabled and two more fields to be added, one of them mandatory.

While that might work in the simple situation of identifying which metadata fields in that second level are mandatory and which are not, what happens when the initial selections not only control which fields must be populated, but the very existence of fields? Consider a situation where a value for one field is filled in, and then another field is filled in with a value that disables the first field? Is the first value retained when it is no longer a part of the current model? And if we revert to the prior configuration, restoring the previously removed field, will the previous value be automatically restored? What would the appropriate behavior be if that value were retained, but it were somehow in conflict with other metadata (e.g. no longer a valid option)?

> **To retain conditional metadata, or not?**
>
> A recent project illustrates this challenge. The project was to manage business-intelligence content for my client's clients. Each client purchased access to a variety of data products and had its own models for categorizing each product.
>
> Our first-level metadata had to identify the client and the data product, which then determined the categorization model for the next level of metadata.
>
> In the next level of metadata, geography was a common axis, but depending on the client and data product, the geography axis might be region, country, state, or something else.
>
> Now, imagine three clients (A, B, and C). A and C have a region axis, B does not. Content is created and populated for A, then someone realizes that the data was assigned to the wrong client. So the client is changed to B, which is again wrong. Finally, it is changed to C.
>
> Initially (for A), the region had a value of France. When client B is selected, we have a stray piece of region metadata; it has no meaning or value. Should it be kept? And when the client is finally changed to C, should the value France be restored? Maybe. It is likely to be correct, except that C uses a different set of values for region; in this case, Western Europe is the correct value; client C's process does not distinguish France as a separate market.
>
> This example suggests that a metadata value should not be retained when it becomes redundant. Of course, if the example were A to B to A, then we might come down on the side of retaining it. On this project, we retained the redundant values, but we also differentiated the conditional axes: A's region field was not the same as C's.
>
> There is no perfect solution to this issue. We end up with stray metadata; values that – based on the configuration of other axes – are never going to interfere with the functioning of our system. But if this will happen frequently enough that excess data could affect performance, then it would make sense to provide a function to clean out excess metadata. Otherwise, retaining a few spare values is probably easiest.

And how do we deal with a scenario where a metadata value is updated in such a way that the categorization pattern changes, leaving mandatory fields unpopulated within published material?

This is where the normal means for handling metadata fall apart. This is where complexity interacts with information, and smart algorithms are required to resolve the holes created by unpopulated mandatory metadata.

How does this relate to author experience? Simply through highlighting the need for those responsible for the information to be prepared to let go; they do not have the resources to control every aspect of the communication across an ever-evolving array of platforms and devices. And even if they have the bodies available to manage so many variants, they cannot control such a sea of data without delegating decision-making authority to those lower down in the pecking order.

## *Solving the metadata challenge*

The issues of manageable metadata can, for the most part, be dealt with through good content design, governance, and author experience design, as discussed in Chapter 6.

# Practical Author Experience

# CHAPTER 5
# *Hierarchy of Author Experience Needs*

In 1943, psychologist Abraham Maslow presented a theory of psychological health predicated on fulfilling innate human needs in priority order, culminating in self-actualization.[1] Maslow's hierarchy of needs starts with the most basic: physiological – elements without which life cannot be sustained. As you climb the pyramid, you find elements that make life more secure, enjoyable, and meaningful.

To follow in these footsteps, I offer you an alternate hierarchy of needs: the hierarchy of author experience needs.

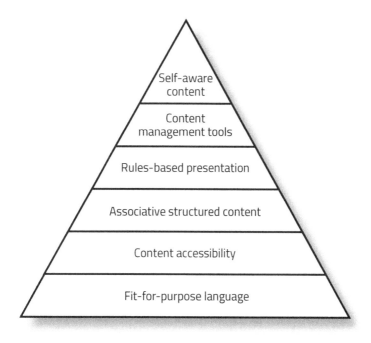

Figure 5.1 – The hierarchy of author experience needs

The ordering of the levels within the hierarchy does not represent any requirement that the levels be implemented in a particular sequence. My ordering starts with the reduction of pain and moves to enhancing productivity and value. The further up the pyramid you climb, the more completely your system fulfills its purpose of facilitating the management

---
[1] From http://en.wikipedia.org/w/index.php?title=Abraham_Maslow&oldid=610238322

of content. The order also maps, to some extent, to some of the challenges to good author experience described in Chapter 4:

- Fit-for-purpose language and content accessibility both deal with the technologists' paradigms.
- Associative structured content and rules-based presentation address the issue of content coupling.
- Content management tools and self-aware content, between them, cover aspects of workflow, content ownership, and mental models.

What follows are descriptions of the elements that make each of these steps a reality. I will admit a bias towards the technical implementation, because each one needs to be implemented.

## Fit-for-purpose language

If I had to guess at the most common response from a client when explaining some element of a system's interface, my money would be on "Eh?"[2] While it's easy to assume the reason is that information management stretches beyond most people's comfort zone, this isn't the root cause. When you explain a process – even a complex one – in terms of a person's paradigms (*domain model*), that person will understand quickly.

The real problems are threefold.

First, people are not used to breaking down their actions, processes, and workflows into steps. They see the whole as a single task and don't clearly distinguish the individual steps.

Second, assumptions about computers are overlaid on expectations for the system. As a result, people focus on particular processes that computers are supposed to be good at. An example of this is the misuse of automated workflows described in the section titled "Workflow that… doesn't."

Third, the terminology used in the interface does not resonate with the audience. This derives directly from the challenges described in the section titled "The technologists' paradigms." The vocabulary used by technologists does not match that used by the audience.

---

[2] At least, my money goes on "Eh?" assuming we overlook the expletive-laced variants. And, in case you get the wrong idea, this refers to not-yet-customized interfaces.

We end up with a disconnect on two levels:

- The actual terminology used
- The conceptual models of access, association, and management

I include the conceptual model disconnect within fit-for-purpose language because we are dealing with how the system communicates with the author. It is vital that when we communicate, we do so in a language that is appropriate to our audience. As Willy Brandt put it so eloquently:

> "If I am selling to you, I speak your language. If I am buying, dann müssen Sie Deutsch sprechen."
>
> —Willy Brandt

## *Not all synonyms are equal*

Let us consider information that has been entered into our system, but is not yet accessible to our audience. What term should we use for the action that switches the state of this information from not accessible to accessible?

- Activate
- Approve
- Enable
- Publish
- Put live
- Release

Each of these describes, to a varying degree, what is happening. But which one describes how the individual performing the action thinks? Even though this is a simplistic example, we still have at least six choices of terminology. The best option depends on the authors.

Or, taking an example from the web, consider changing the URL of some web content. Does it make sense to use the term *move*, when this node in our content structure is referred to as the *page name*?[3] Wouldn't the term *rename* make more sense?

Or when describing the purpose of content, we might ask: how many interchangeable terms are there? Is it a title or a heading? Does it need an additional descriptor? Do we need to distinguish between the title that describes an article as an entity and a variant used to cross-reference

---

[3] In this case, it is clear that the terminology derives from a developer's past, working with a Unix system where renaming a file was described as moving it and was effected by using the mv command.

it from some other content? What term best enables the author to understand what a particular element is, what it does, or how it works?

> **Counting titles**
>
> One intranet project I was recently involved in had six title fields for every block of content:
> - The name of the content element
> - A main title
> - A title to appear in the browser header/tab
> - A navigation menu title
> - A breadcrumb title
> - A teaser title for cross-references
>
> Each of these serves a purpose based on context, design, and navigation.[4]

There is no single answer. The terms that work in a particular situation depend on the business, its domain models, the industry, the authors' mental processes, the terminology and jargon within the organization, and more.

Given the dependence on authors' experiences and prejudices, there isn't always a natural consensus. Someone may feel that another term could be better. What matters is that terminology must come primarily from authors rather than developers. Terminology must align with the author's experience, not some technical process.

## *Explaining the obscure*

Content management often requires authors to add information that they understand in principle, but have no meaningful name for. Maybe they can come up with a name; maybe not. But it is unlikely such a term will be meaningful the next day.

In this case, the most easily understood terminology would be a verbose description. Consider the part of a URI that identifies an instance of content for access. For example, `sample-article` in the URI `http://domain.com/articles/sample-article`.

In some systems, this is referred to as a *slug*. Now, I associate the word slug with a shell-less mollusk, the projectile used in some weapons, or

---

[4] The reason all these were needed is complicated, resulting largely from visual design decisions made before the content was fully understood.

a blank piece of material that will be machined into a final product. With a little research, I find that journalists use slug to refer to a unique identifier key for published articles, so I can appreciate the evolution to a web-centric form. But is the term really understood? With repeated use, maybe, but it is still hard to disassociate the mollusk.

Personally, I prefer a term like *content identifier,* but neither of these terms is commonly understood, so it doesn't particularly matter which is adopted in the system; the authors will learn. But where such a learning curve exists, you must make help easily available. That might be something as simple as associating explanatory text with the element, as shown in Figure 5.2, so new authors can understand what the field is for, and why this obscure element is mandatory!

Figure 5.2 – Explaining the slug

## *Consistency of terminology*

In addition to being suitable for the authoring audience, language must be used consistently throughout the interface.

Do not call a button Activate and then present a confirmation message saying "Your content has been published." While many use the terms interchangeably, and you probably have no trouble understanding that they are synonymous in this case, the disparity will confuse some.[5]

How would you fare if the terms "dát živě" and "povolen" were used interchangeably? Your Czech isn't that good? How, then, can we expect those for whom English is not their mother tongue to understand the fascination we have with using synonyms instead of repeating terms in consecutive sentences?[6]

This principle of consistency extends beyond the call to action and associated confirmation.

---

[5] As well as being confusing, it also demonstrates sloppiness – blatant disrespect for the author as a user of the system.

[6] I'll be the first to admit that I am guilty of avoiding repeated terminology between phrases, particularly in prose. To any readers who are not native English speakers and are struggling with my use of language, I apologize.

If you repeat elements between different content types, use consistent language. If *title* is the top-most element in one type of information, it makes no sense to identify the equivalent field as *heading* elsewhere. (Do not confuse title and heading here with the HTML elements `<title>`, `<h1>`, `<h2>`, etc. I refer to their semantic values in context, not what they represent in a specific output.)

That last statement must come with a disclaimer: just because an article has a title does not mean that every content type has a title. Within an article, you have a title and a body, and within the body there is further structure; for example, sections likely have headings. As such, if you have elements of information managed at a granular level that equate to reusable paragraphs, they have – in this example – a heading rather than a title.

The purpose of consistency is to reduce the learning curve – not only the time it takes authors to learn the system initially, but also the time it takes when new functionality or elements are added.

> Authors' reactions when a new element is introduced should be: "How did I miss that?" rather than "How does this thing work?"

## *Consistency of management logic*

In addition to using consistent terminology for content, we also need to use consistent structures. Just as body language and tone convey information in spoken conversation, so the structure of our content types is a language that conveys information about hierarchy and priority.

Imagine two types of content: an article and a person profile. Each can have images associated with it: the article has a lead image and images within the body; the person profile has a head shot. Depending on how the information renders, there may be a need for variants.

What attributes will each image definition have?

- A reference to an image managed in a library? Optional, but if populated, this reference may override the availability or characteristics of other fields.[7]
- A binary file? Obviously.
- A name, for the purposes of in-system identification? Certainly.

---

[7] This is a classic case of metadata on one element changing the metadata requirements for other elements.

- Alternative text? Mandatory for the article-body image but optional (self-populating) or even non-editable for the head shot and lead image.
- A caption? Probably. But the format of each may be different (for example, formatted text for the article-body image, but only plain text for the article and person profile).
- Rendering options? Possibly, but not necessarily the same for each image type.
- A linking reference? Only for the article-body image.
- Associated paragraph? Maybe, but only for the article-body image.

The result might end up looking like Table 5.1.

Table 5.1 – Configuring attributes for an image

| Attribute | Person-profile | Article | Body image |
|---|---|---|---|
| Reference asset | y (opt) | y (opt) | y (opt) |
| Binary file | y | y | y |
| In-system name | y | y | y |
| Alt text | n | y (opt) | y |
| Caption | | | |
| Plain text | y (opt) | n | n |
| Rich text | n | n | y |
| Rendering options | | | |
| 4 × 5 | y | n | n |
| 1 × 1 | y | y | y |
| 3 × 2 | n | n | y |
| 16 × 9 | n | y | y |
| Link reference | n | n | y (opt) |
| Associated paragraph | n | n | y |

With such variance, how do we group these attributes?

While the easy answer is to say "include them all sequentially," we need to consider how complicated some elements might be if we did that. I have seen cases where clients went overboard and defined elements whose attributes would extend to three screens if listed sequentially. In such cases, the attributes need to be grouped and clustered using a tabbed interface, collapsible sections, or both.

In practice, the best option is to list the full set of attributes that need to be configured for the content type, as shown in Table 5.1. Then, we can spot patterns. What groups do all variations have? Which attributes are specific to only one or two variants? In this case, I would start with three tabs: the basic information (reference, file, alt text, caption), the rendering options, and references. The references tab would only be present for the body image variant. While this is a minimal example, it should give you a sense of the possibilities.

It is also worth keeping in mind that the best option is often simplicity. Why does the design demand so many rendering options?

We need to accommodate attribute variance so that when an author encounters a new arrangement, the layout and grouping make intuitive sense. I don't claim that the structure in Table 5.1 is the only correct structure for this case. There are other equally meaningful ways these fields can be organized.

What matters is that the associative language of the elements – the mental model – is appropriate to the authoring audience's needs and their concept of the image. There is no single right answer.

> **Structural definitions that work… to a point**
>
> Over a range of web-only projects where mobile/responsive design was actively out of scope, I found that for most types of content, a content/presentation split was generally the most easily understood. However, the model was tenuous at best (at least half the content types needed a third category), and it clearly fails when we need multi-channel delivery.
>
> A more recent project largely solved these problems with a better up-front definition of the semantic meaning of content that reduced the options. By placing image reuse out of scope and restricting how images could be used, I ended up with the options shown in Table 5.2.
>
> With this simplified attribute set, the grouping is easy.

Table 5.2 – A reduced set of attributes for images

| Attribute | Person-profile | Article | Body image |
|---|---|---|---|
| Binary file (in pixels) | y (140 × 140) | 2× (631 × 390 & 195 × 121) | 2× (1600 × 989 & 390 × 241) |
| Alt text | n (auto) | n (auto) | y |
| Caption (text) | n | n | y |

## *Content association paradigms*

How can we best create, represent, and maintain associations between different elements of information?

Most people's answer, based on experience of how these links are presented within the web, is to use the HTML anchor tag with a hyperlink reference `<a href="...">`. This format tells a browser to render a link to the referenced content. The mindset is so common that an embedded link is one of the most common ways of making such an association.

There are two problems with this.

First, the author needs to find the page to link to, which usually means browsing through some form of content tree (all too often in an under-sized window).

The second, and more important, problem is that this couples content with function. Such a link creates a hard dependency and an implicit assumption that the primary output is a good enough representation. If the target is removed, revised, or split, the reference becomes invalid, incomplete, or broken (and no one likes broken links).

Using a hard link to create an association does not come from any natural human behavior. It is just the simplest technical approach to a complex problem. A hard link solves one instance of the problem (I want to create a link between A and B) without considering the complete process (I want to create and maintain relevant associations between A and all other elements with which it shares a subject affinity, without incurring undue overhead).

There are cases where a simple one-to-one association is appropriate. One example is when we want to associate an image from an asset library with a person's profile. We want a specific image for that purpose, not just any image that matches the identified individual. But even then, we need the reference to be maintained so that the system can cope with the target being removed.[8]

However, don't let the edge case of a single association be the guiding paradigm that dictates all associations. Instead, let the general case guide the author experience. After all, the general easily handles the specific.

The correct tool for creating associations between content is reference metadata: attributes that establish subject affinity. This way, the system can find all relevant content,[9] whether that content is a single element or a set of hundreds. The system then handles the reference by applying a set of rules to the results it finds.

And if you want to create a reference to a specific asset or piece of content, a metadata filtering model is also an obvious way to reduce the target list to a manageable set of candidates.

## Content accessibility

In the section titled "Content association paradigms," I outlined the problems that can occur when a content management system has the wrong interface for creating associative links. The same interface challenges occur when you are looking for material to edit in the first place.

The scope of this issue depends on three factors:
- The size of the content store
- Whether the content is intended for one output or more than one
- Whether the content is page-centric or granular

A small content store in a page-centric model with only one output (and no consideration of expansion) is the easiest.

---

[8] This possibility also dictates how you craft copy that surrounds the target, because the supposed linking element may not always be a valid link.

[9] In his book *Every Page is Page One* (XML Press, 2013), Mark Baker covers subject affinity in detail, explaining how semantic identification of elements within copy can be used to provide cues to a system that is designed to look for and inject related links.

## Hierarchy of Author Experience Needs 63

As we enhance the system, things get messier. More content demands more structure. Granular content increases complexity. Multiple outputs demand a degree of abstraction in the model. This last point is especially important when some outputs are adaptive, without *page parity*.

There are four basic models that we can use to access stored content:
- Output-based navigation
- The tree model
- Filtered types
- Dynamic multi-axis content filtering

Each model in the list is more difficult to implement than the one above it, but more capable of coping with the complexity of a larger store, more outputs, and granular elements. Note that these models do not reflect how content is stored. They represent the navigation used to access the content. And they are not mutually exclusive.

### *Output-based navigation*

With output-based navigation, you use the same URI to retrieve content for editing that an end-user uses in the published environment. But this is only part of the model. To provide output-based navigation, the published environment needs to be presented within the author environment, since it is used as the means of navigating to the content. This is similar to a blog system, where a logged-in user is presented with an Edit link that provides direct access to posts for editing.

Where content is granular – where presentation results from the amalgamation of reusable elements – this model can only take you part of the way to accessing content for editing. From the referenced output, you still need to access the actual element you want to edit.

Perhaps the greatest weakness in this model is that any data stream that feeds the output – content that does not have a URI presented in the user's browser – will be very hard to find.

### *The tree model*

The tree model is often closely associated with output-based navigation systems. Many systems that use tree models map those tree structures directly to the URIs used to access content.

This model divides content into categories and may provide further layers of categorization to subdivide the set into manageable blocks.

Common categorizations include type – products versus articles versus news – date, and language.

While this model can work in some cases with limited content, it suffers under the strain of large content sets because a tree forces a hierarchy of categorization attributes. With a tree model, each element of content has exactly one home, one way of navigating to it. Is the support content for a product within the scope of product information or support information? While both routes make sense, this model requires you to select one as dominant. And any such prioritization is inherently arbitrary.

## *Filtered types*

Once we get beyond the tree model, to structured data, we start towards a model that is viable for large, complex content sets. But many systems that have gone down this route stop short at a halfway solution that is potentially counterproductive, though not always inappropriate.

Many systems enable authors to find content based on metadata but inject one anachronism into the process: first, authors must identify the type of information being searched for: an image, a file, an article, a product, a press release… This is an understandable distinction: it feels appropriate because the available filtering could depend on the type of information being accessed. (An annual report, for example, has very particular attributes, such as a fiscal year. An image, on the other hand, is identifiable by a creator, a source, licensing model, subject matter, theme, etc.) The filtering attributes relevant to one content type may be meaningless to others.

The downside of this model is that it requires authors to know what form of information they are looking for. They need to know the entire information store. Effectively, they need to know where to find the unknown content before searching for it.

In reality, if there are two related elements of different types, finding one could easily hide the other.

Injecting that one level of tree structure into the metadata-based, content-filtering approach, thereby creating a hybrid, might sound like a rational thing to do. However, it is done not because it is good for authors, but because it is significantly easier than implementing full dynamic multi-axis content filtering. When the content type is preselected, filtering can be performed on a rigidly defined set of metadata axes, rather than across multiple, potentially irrelevant axes.

## Dynamic multi-axis content filtering

The pinnacle of content accessibility – at least where a huge pool of content is concerned – is a free-form, search-like approach that surfaces everything that matches the terms used and has the ability to provide context after the fact.

Rather than forcing a single filter, as we saw with filtered types, dynamic multi-axis content filtering accepts a search value, then offers filters, but only filters that are valid in the context of that search value.

If you are searching for an element of content and enter the value 2012, what are you referring to? Is it a product in the 2012 catalog or one that was released that year? Is it some article concerning your company's marketing efforts around the London 2012 Olympics? Is it corporate financials for 2011-12 or 2012-13? Is it a departmental report written in 2012 or a projection relating to 2012 but written years earlier? Could you, perhaps, be searching for the article about the 2012 people who attended an event (because you happen to remember the number)?

We need an interface that surfaces relevant, and only relevant, filtering axes. What such an interface would look like depends on how, where, and by whom it is being used.

A format I have seen some (technical) people recommend is a form of pseudo-code, a format that makes some sense in its own right. This might be something along the lines of [axis-name]:[value]. These people also suggest using Boolean logic operators (AND and OR) and parentheses in search queries.

While this coded approach may allow the query to be easily parsed by the system serving it, it is unsuitable for most authors. Rather than having the system select the filtering axes, this approach forces the author to enter them. Once again, the system's inner workings and terminology are surfaced. How many authors know the technical name of the axes they want to search on? Or have a clue as to the range of axes available to search against? How many will feel comfortable with the syntax?

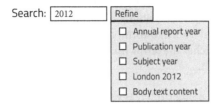

Figure 5.3 – Refining search

There is no perfect solution here, but I always come down on the side of getting the system to do most of the work. In this case, after 2012 is entered, why not have the system translate the options into something the author can understand and use to make a meaningful selection. This could be a control button next to the search box that displays the axes that the search value could be filtered against.

Figure 5.3 shows what such a control button might look like. Obviously, the drop-down should only offer axes that will return results, and it must display options in an order that maintains a balance between the weight given the most relevant axes and consistency. This approach demands that someone in the organization be responsible for maintaining the metadata and search refinement axes.

## Associative structured content

The content coupling problems discussed in the section titled "Content coupling" – that is, content being coupled with elements of presentation, context, or behavior – need to be dealt with through a combination of approaches. First, content storage needs to support separation of these elements from content. And second, the author interfaces need to make it possible to manage the content and these elements without the process being overly onerous for content managers.

There are two parts to separating the elements: breaking the content out from the presentational, contextual, and behavioral coupling, and pulling it back together within the output environment (covered in the section titled "Rules-based presentation"). When we separate content from all coupling, we need meaningful associations to retain meaning.

## Constraints

Many people are of the opinion that the greatest empowerment comes from the most flexible tools. They want all the utilities and features available, so they can make the best choice for the particular situation.

With reusable content, however, authors cannot know the context in which their material will be consumed; they do not even know the platform it will be delivered to. As such, the only power that comes from flexibility is the power to make an even bigger mess of things.

Lack of knowledge about output means that structures must be consistent. For that, we need limits. These constraints come in three forms:

- What must be included (mandatory fields).
- Flexibility that must not be made available (most of the standard WYSIWYG toolbox).
- Types of content that can be created (the scope of content the system is intended to manage).

These constraints create value in many ways. They ensure consistency, which plays into branding. They encourage authors to pay attention to crafting the message body, so that it expresses its meaning through the words used rather than situation-specific presentation; this has the secondary benefit of making the message more accessible. Limiting types of content ensures that the content aligns with the business purpose of the system.[10]

Professional authors do not complain about such limits; they already operate in this manner, whether the constraints are enforced by the system or by their ethos. It is amateur authors who raise the greatest ruckus, and it is they who would benefit most.

## Structured content

The idea of structured types has been around for a long time. Anything that can be described using concepts rather than specific instances is a structured type. The concept of a car is a structured type; the car you drive is an instance of this type. In computing, structured types are

---

[10] Limiting content types – and content instances – is based on the principle of the Content Validation Model outlined by Rahel Bailie (@rahelab on Twitter) at Content Strategy Applied in 2012. I subsequently wrote up details of the model, which can be found at: http://think-info.com/2012/04/16/the-content-testing-ground/. Basically, goals are defined for the system; objectives are defined that fulfill these goals; end-user-actions support the objectives; and all content must directly enable actions. Any content that does not have a defined place within this validation model does not belong in your system.

fundamental to the concept of object-oriented programming, which is central to many of the more popular languages in use today.

Structured content is "content, whether in a textual, visual, or playable format, that conforms to structural and semantic rules that allow machine processing to meet specific business requirements."[11]

A structured content type has mandatory fields that are fundamental to the definition of an instance and, possibly, optional attributes. It follows a pattern; it is recognizable. Without structured content, it is impossible for content within your CMS to be used across platforms and channels. Without defined structure, there can be no reusable rules for presentation or behavior.

## *Reusability and repurposability*

One of the first times I talked about author experience in public, I used the metaphor of cake.[12]

If I were to place several cakes before you and ask you to choose the one you want, you might be able to make an easy choice, or you might dither because two or three appeal. (For this example, I use two cakes: ginger cupcakes and three-chocolate brownies.) In some cases, flavors or texture dictate your choice. In others, it is a matter of presentation. It doesn't matter; there is, effectively, one preferred cake that is optimal for you.

So long as I have all the cakes, I can serve you the one you want.

If I had only cupcakes and you wanted brownies, we would have a problem. Maybe you would eat a cupcake anyway, but you would feel cheated. You missed out on what you really wanted.

Of course, there is a problem with making several sets of cake: the cakes people want may not be the cakes I have available.

But what are the difference between cakes?

They are all based on the same core ingredients, albeit in slightly different proportions. Some have an additional condiment or two that the others lack. Fundamentally, they are all cake. So, instead of having several cakes, I could have one set of ingredients that covers all the recipes. What I then have available to display is not as appealing as the finished product

---

[11] Definition from *The Language of Content Strategy*, (XML Press, 2014).

[12] It was, after all, a content strategy conference, and content strategists have a purported affinity for cake, especially cupcakes.

– the ingredients do not titillate the taste buds the way that seeing, and smelling, cake does – but I have flexibility to serve you what you want. Not that I am going to show you the ingredients per se, but you can request the cake that best suits your needs, and I will make it.

It might take a while to produce the specific output you require, but in the digital world, a computer is doing the mixing: it does not need to bake for 45 minutes.

The point is that you do not need the sum of all the ingredients from each of the recipes to bake whichever cake is requested; you need only the superset of ingredients that enable you to bake any of them. (In the digital world, using something does not destroy it, so you can bake to your heart's content.)[13]

But if any of the ingredients are already mixed – partially baked – they can no longer be easily reused.

This shows the value of retaining the elements that make up our content as individual ingredients, unbaked. To add an extra cake to our offering, we require, at most, a few additional items not already covered by our initial recipes and new instructions for mixing. We have already established that cakes are, for the most part, the same basic set of ingredients. This is repurposability; it is what makes structured content so valuable.

## *Semantic relationships*

There is a tendency within many systems to create blocks of content that are usable as-is. This occurs in many forms, one of the most basic being the definition of associations between elements. There are two parts to this: the element we create a reference to (the target) and the bit of content that contains the reference. To refer to a person, we use a name. For a media asset, we embed a link to it. If we identify a book, we might link to it on Amazon. But in doing so, we create hard structures, rather than associative ones.

With the Amazon example, which site are we linking to? You may be in the US, where .com is appropriate, but I am in the UK, so I consider the .co.uk link more relevant. Amazon does not use the user's location and login to redirect a link to the product page on the local version of their website. To make matters worse, I do not know if you want the hardback, paperback, or Kindle version. (Which raises the possibility that you have a different e-reader or prefer to buy from someone else.)

---

[13] As a bonus, digital cake is free of calories!

These variations are bad enough, but what happens when the book is withdrawn? Now, my link – whichever variant I chose – is broken.

Instead of embedding a link within the content, I should use a reference to identify the book semantically. That reference can then be rendered appropriately for you.

This associative structured approach has benefits beyond simply serving the best link. It encourages us to think about the best source for the linking text. Because we have an association with the content we are linking to, it makes sense to obtain the linking text from the link target, so we can refer to it by its correct name. If the target's name changes, obtaining the link text by reference keeps it correct.

> **Book linking on Wikipedia**
>
> Wikipedia provides us with a partial example of the marker instead of the hard link. Unfortunately, it does not do anything useful or user-friendly with the associations it creates.
>
> If you reference a book in Wikipedia, you use its ISBN. To stay neutral, you do not create a link to Amazon or any other bookseller. Instead, the link takes the user to an intermediate Book Sources page that links to the book in various databases, general search sites, libraries, book swapping sites, and booksellers.
>
> There are two problems with Wikipedia's approach:
> - No checks are performed on the linked-to databases and other sources to determine whether the book is referenced there, so many of the provided links are dead ends.
> - The full list of libraries is displayed without taking the user's location into consideration, so the page provides an overload of links, most of which are irrelevant.

The associative structured approach offers significant benefits from a governance perspective. What happens when a person changes his or her name? Without semantic referencing, how are we supposed to find every reference to the old name and update it without error? Or where our referencing strategy includes social media cross-references, and a person changes his or her online identity, how is the author supposed to track down all instances and make hundreds of updates?

Associative structures enable the system to do the heavy lifting, a task for which a digital system is eminently suited.

## *Dynamic associations*

One scenario where the use of semantic relationships is relevant is the inclusion of media within other material; for example, an article with an associated image or video. If that media element is licensed, we run into governance issues when the license expires. Obviously, we need to be able to track down all the places where the media is used. But what do we do about replacing it? What if the image fills a mandatory role?

While we could require each instance to be updated manually, this is risky. Instead, we need fallback mechanisms that automatically select alternate media. There are two basic approaches.

The first approach is to manually configure fallback options: populate multiple candidates. However, this only defers the point of failure.

The second, better approach is to reference the desired element through content filtering. Of course, we can identify a preferred asset, but the system can use metadata to adjust the content when things change, reducing the possibility of invalid references. The rules that select any particular asset may differ, and when circumstances change, the system may choose a replacement asset. The association becomes dynamic.

## *Sequence and narrative flow*

When we present information to our audience, a large proportion of that information is usually in the form of sequential narrative. This is straightforward when dealing only with words. Just string them together in paragraphs with the occasional heading to supply context. But when we are dealing with multiple media, we have more choices; we can include imagery, slideshows, videos, call-out highlights, asides, or more complex interactive elements such as dynamic maps or animated models.

If our narrative is to be presented across media – able to be consumed as audio, print, web, or other media – we need the structure of these elements to make sense in the authoring environment as well as when presented. We need our systems to deal with the positional layout of print, the dynamic layout of the web, and the hub-and-spoke model suited to mobile.[14]

---

[14] The hub-and-spoke model is not only relevant as a means of accessing rich media within a narrative flow; it is also a dominant paradigm within transactional interaction. Using a structured content approach solves the functionality coupling problem too. For more, see *Rethinking Mobile Checkout* [http://www.lukew.com/presos/preso.asp?30], by Luke Wroblewski.

It is fundamentally impossible to create reusable complex narrative content of this sort without structured content, separated from presentation.

## *Information hierarchy*

If the telling of a story varies depending on the medium, how can we craft reusable content?

When creating a single instance of a story for a single platform, using an appropriate visual semantic language (formatting that aids understanding) is easy enough. The important points are made prominent. Relationships between blocks are handled through different weights of headings to express a hierarchy (just like the headings in this book).

The problem is that while we can easily express these relationships using visual semantics, which most people understand intuitively or can learn quickly, expressing these relationships through descriptive identifiers is difficult for many people.[15] The solution is to have tools that surface a limited, generic, semantic formatting language to the author and that do not allow authors to create new definitions: all relationships are represented with context-specific, abstract semantics.[16]

In a situation like this, there are levels of document structure that tools need to handle that are not directly displayed to the author, but that are inherently understood when using visual semantics.

For example, if we have a chapter in a web-published document, we would expect the mark-up to look like Example 5.1.

---

[15] To understand the difficulty of using descriptive identifiers versus employing a visual semantic language, consider the use of word processors. Despite styles having been available for more than a decade (two if we consider certain tools), many people still mark up their documents manually with bold, italic, and font sizes to express their information hierarchy. The push by the likes of Microsoft to make styles more prominent may have influenced some people, but semantic definition has yet to achieve mass traction.

[16] Yes, I am well aware that "context-specific abstract" sounds like an oxymoron. It is not. The abstraction is in the presentation. Headings are identified by the semantic levels they represent: titles, parts, chapters, sections, etc. The context-specific part indicates that the semantic language available is limited to the particular needs of the information being worked on.

Example 5.1 – Web-specific chapter markup

```
<p>Paragraph before chapter.</p>
<h2>Title of chapter</h2>
<p>First paragraph of chapter.</p>
<p>Second paragraph of chapter.</p>
<h3>Section title within chapter</h3>
<p>Paragraph within chapter section.</p>
<h2>Title of next chapter</h2>
```

But the semantic structure we understand when we see the output is more like Example 5.2:

Example 5.2 – Semantic chapter markup

```
...
  <para>Paragraph before chapter.</para>
</chapter>
<chapter>
  <chapter-title>Title of chapter</chapter-title>
  <chapter-body>
    <chapter-intro>
      <para>First paragraph of chapter.</para>
      <para>Second paragraph of chapter.</para>
    </chapter-intro>
    <chapter-section>
      <chapter-section-heading>
        Section title within chapter
      </chapter-section-heading>
      <chapter-section-body>
        <para>Paragraph within chapter section.</para>
      </chapter-section-body>
    </chapter-section>
  </chapter-body>
</chapter>
<chapter>
  <chapter-title>Title of next chapter</chapter-title>
...
```

The visual elements tell us a lot about the information, including details that a computer system can only understand if we give it additional processing rules. In this case, Example 5.2 can easily be derived from Example 5.1, but it would be much better if the system did not need to derive this information from a presentational structure. And there are scenarios that have no easy visual representation.[17]

You may notice that I changed the tags. I used `<chapter-title>` and `<chapter-section-heading>` instead of `<h2>` and `<h3>` respectively. I used `<para>` instead of the easier `<p>`. These changes are there because the more verbose identifiers are structural; they are presentation-format agnostic. They tell us (and the computer) the semantic value of the information they contain, with no information pertaining to display.

If we broke this sample book for rendering into one page per chapter or section, then the `<chapter-title>` or `<chapter-section-heading>` elements, respectively, would need to become `<h1>` for web display, since they would be the respective page-level titles. If we embed any one presentation format into the storage, we undermine even this simple reuse example.

## *Emphasis and proximity*

Semantic structuring of information within the system provides other benefits. If an image is required, we can choose how to include it. Does it belong at an exact location within the flow of the story – behaving as a paragraph within the narrative – or does it have a subsidiary relationship to a chapter, section, paragraph, or even a cluster of words?

Many CMSes treat the image as a block. It belongs between paragraphs. This is the easiest to implement from a technical perspective, and it does not require authors to mess around in code.

But if we associate the image with a block (as metadata), then we could adapt the display based on the platform and channel using simple rules. For example, if we associate the image with text, say "in Figure 1," then once the system has positioned the image for a particular output channel, it can generate a proximity/direction modifier (link, below, right, opposite) with the reference (e.g., "in Figure 1 below").

An information storage model that includes a full semantic model[18] and associative definition enables the system to represent the visual semantics of emphasis, proximity, and hierarchy automatically, without forcing the author to create individual adaptations.

---

[17] Please, tell me how you would show a paragraph at the end of the chapter, which is not part of the chapter-section that precedes it. A wider margin between paragraphs, perhaps? But how do you identify that break point in the presentation-only code?

[18] No, I am not suggesting we surface the code-like fragment I showed above to authors.

# Rules-based presentation

In the previous section, I argued in favor of creating content based on structured types with semantic associations and a computer-readable representation of hierarchy and relationships.

If we break our content out of an arbitrary container and structure it semantically, we have what appears to be a problem but is in reality an opportunity. There is no predefined way of displaying this content, no default rendering.

This is a Good Thing™.

## *Adaptive presentation*

When content exists in the abstract, the presentation playing field is leveled. No preeminence is given to web or mobile or print output. All forms are equal. All displays receive the same consideration.[19] And when each format receives its due consideration, everything works better.

The current trend in digital publishing is toward using responsive web design (RWD). The theory of RWD is that you can serve a single version of the content, regardless of context, and use device detection mechanisms to modify the presentation. The theory is that you do not need different outputs for desktop browser, tablet, and phone.

My issue with RWD is the assumption that the web is preeminent. Yes, digital media have the greatest potential, and we can expect that, for the foreseeable future, digital carriers will be involved in ever more of our communications. But we must not focus so intently on the web – whether desktop or mobile – that we lose track of everything else.

Web and digital are not synonyms.

Fundamentally, RWD considers the browser to be the only delivery medium/channel. It reduces the diversification of platforms to a single responsive mapping and assumes content can be optimized for that default channel. In effect, it ties itself to the idea that there is a primary delivery medium.

---

[19] You can argue that some displays are "more equal." But if a display has a large enough audience to warrant being catered to, that audience deserves proper consideration.

This is not the case. There are many delivery channels, and RWD can serve only a narrow band. We need a solution that can present our content to all channels equally, without onerous overhead.

RWD assumes that delivered content can be divided into chunks of a certain size that will be equally good for everyone. In truth, we all consume information in different sized bites. Where one channel might be best served by a single stream of content, with defined placement of media within its flow, another channel might be better served with a multistep paginated model, perhaps with a menu to jump to a specific location within the whole, with media as separate entities that must be requested. Figure 5.4 shows this contrast between optimized print and mobile models.

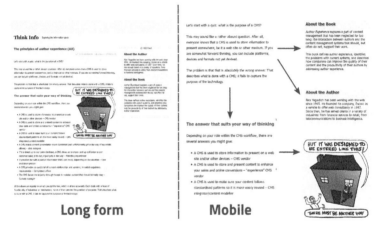

**Long form** | **Mobile**

Figure 5.4 – Long-form article vs. mobile pagination

Some channels demand a completely different set of presentation rules. They may use different mark-up languages to define content flow. They may be data streams rather than renderable pages. They may use some output form that we cannot yet envision because the channel will not be released to the public until next year.

To push content to all those channels, we need to adapt our output to the needs of the channel, rather than relying on one all-encompassing format that responds to the rendering environment. We need to do the work when we serve the content; we need to separate it into logically-sized bites suitable to the delivery medium. This will become very useful when we finally have a mechanism for detecting the status of the client's connection, so we can ensure that we do not try to push huge files to a user with a slow – or pay-by-usage – connection unless that user actively requests them.

And if we have the technology to distinguish between web, app, print, and whatever comes next, shouldn't we use that technology to differentiate how we serve web content, adapting to what we know about the target device? Why mix technologies?

This approach of serving up rules-optimized content directly and not relying on the user's environment to vary it is *adaptive content delivery*.

Adaptive content delivery is balanced. It does not prioritize one channel over another. And it is surprisingly simple to do if you take a structured and template-based approach. Unfortunately, providing a full technical description of how this is done would double the length of this book and change its tone.[20]

## Content for the user's context

The idea of adapting content must not be limited to consideration of the device it is being served to. Indeed, that is a relatively new area. The older and better established definition of content adaptation is serving up completely different content based on the user's context, to the extent that we know it.

I include this discussion here for the sake of completeness, but also because the RWD approach runs the risk of confusing people into thinking there is a one-size-fits-all solution, and because, when we are dealing with adaptation for both device and user context, there is a risk of complexity being used as an argument to remove functionality.

For example, when selling a product, the context of users looking at that product for the first time is distinctly different from the context of those who have already purchased the product. We can determine what secondary information is most likely to be relevant to each group – maybe comparison tools versus support and accessories. Of course, the contextual displays would need to be tested and optimized.

## Contextual presentation models

It is fairly easy to understand the immediate value of not considering one presentation format to be the default for a specific content type. With a level playing field, the merits and needs of the web, a mobile app,

---

[20] A good starting point on adaptive content and why it can't wait is Noz Urbina's recent post: http://urbinaconsulting.com/2014/09/11/adaptivecontent/

a video/audio feed, print, and data feeds are all given due consideration. Each channel has its own rules.

But there is a hidden benefit to this model.

The outputs we have discussed so far are all self-contained whole output models: the article in its entirety. There are three further scenarios that we need to develop presentation rules for:

1. Partial renderings of a content instance. For example, you might have a reference to an article that retrieves just the article title and author, then displays those two pieces of information in line.
2. Aggregations of multiple instances or partial instances. For example, a list that contains the title and author for each article in a group of related articles.
3. Rendering content that has been changed in a way that invalidates its structure (as in the section titled "Meta-metadata").

It will not always be relevant or meaningful to deal with all three scenarios for every medium, but we need to consider them.

Additionally, there are cases where multiple variants will be required for the same medium. Imagine an article presented on a web site. We may need a list of links with items such as the following:

- An option to feature the article in a hub or landing page
- A link in a list of articles
- A cross-reference link from another article
- An inline text link
- A search result representation

These are just a few possibilities. Depending on the specific context, there will surely be others. Each needs a presentation model based on the content structure and rules.

## *The input presentation context*

Since we have been dealing with contextually adaptive presentation of content types, the thought may have crossed your mind that the authoring interface for a content type is little more than another rendering of it that happens to support writing back changes.

To this I say: absolutely.

Taking this view means that we can have multiple authoring interfaces for a single content type, depending on author needs. And if the model is appropriately designed, these are as easy to configure as a new device output structure.

## Adaptive previewing

Using structured content and rules-based presentation means that the input model for our content is going to be predominantly forms-based.

So, returning to one of our earlier themes regarding content, structure, and presentation, if the interface is forms-based, how are authors going to make the mental connection between the content they are managing and the way it reads when presented? After all, we read the whole message, which includes the presentation.

The answer to this has two parts:

- The first part is to create a forms-based authoring environment that does not feel like a form. I cover this later in the section titled "To form or not to form?"
- The second part is to provide authors with a realistic, indicative preview of the output variations.[21]

The area I want to deal with here is the second: the available previews. It does not matter what kind of CMS we are dealing with: this requires additional development effort.

With a web CMS, the system generally covers end-to-end creation and publishing flow. The same system supports authoring and publishes content. With a tech comm CMS, and some others, there is a clear distinction between the input environment and the output. It may seem that the web CMS already has this solved. However, a web CMS only addresses one part of the requirement. In addition, as we know, the idea that the web version of content is primary is misleading.

While a web CMS can render a sample output for the web, it does not offer a native mobile app preview within the authoring environment. The native app in most cases accepts a data feed that it manipulates and then presents based on its internal rendering and behavior rules. We need an emulation layer within the CMS that duplicates this app-side manipulation and then previews the output.

And we are not talking only about mobile apps. We need this same preview model for all output channels, from print to screen to…

---

[21] Ironically, we inevitably find that once authors use the system long enough, they create mental mappings that enable them to visualize the outputs without checking the previews.

This means that the interface designers who develop output formats across channels need to do more work; they need to create previewers for each specific format. And it must become a part of their process to update the previewers whenever there are changes to the way a channel presents content.

An alternative to in-CMS previewing is to have a separate publication-QA environment dedicated to testing. Both approaches have costs. Which is better for you depends on your situation.

## *Behavioral previewing*

Information is not useful if it sits, untouched and unchanged, in some repository. It has value only when it is exchanged and when it adjusts and adapts to context. Information is perhaps the ultimate transactional currency.[22]

The key point is that we need to interact with our information. Regardless of how it unfolds in a particular environment, information's *what* is at the mercy of its *how*.

Interaction behaviors provide the exchanged information with context. As such, it is not enough to preview a static display of content across various formats; we also need the authors to be able to safely test interactions and the behaviors that result from those interactions.

The obvious behavioral preview is the transaction: do purchasing systems work properly? There are two approaches to testing this:

- Set up payment methods (such as fake card numbers) that let you mimic the interaction using valid and invalid inputs.
- Add a flag to the data, identifying it as a test that originated in the author environment and allow it to run its full course.

The key difference between these approaches is that the former tests only that the transaction is successfully initiated; the latter tests the full process. That is, does the order get into the system correctly, and does the consequent workflow work? Of course, the flagged approach requires that all parts of the system know how to cope with test data, so the right approach depends on the situation.

But the transaction is not the only type of behavioral previewing. Transaction processing is a technical implementation, rather than a content one. It is unlikely to be subject to much change.

---

[22] A grandiose perspective, perhaps, but one that makes an important point.

A more important aspect of behavioral previewing is to handle the automatic spread of content. If you create an article, how and where do the links to it integrate into existing content? What are the contextual rules that make it appear in various places? How does it stack up against other, similar content?

And how does our article behave when adapted from a single long-form page into a hub-and-spoke form and presented in a small format?

Or if we have two new content sets (in this case, think of a content set as a new product and its associated elements) we are preparing to release, how would the current content behave if only one or the other were to be published? What will happen when both are made available? And what are the consequences of removing content?

> **Concurrent content preparation; staggered release**
>
> I worked with a client who faced this issue repeatedly. Their web home page had a revolving stream of promotions, either special offers or new releases within one or another of their product sets.
>
> The standalone elements for each promotion were easy to manage in isolation. But when it came to managing conflicts on the home page, their answer was "we talk to each other."
>
> This meant the home page content that highlighted or linked to a promotion could only be authored at the last minute, once everything else had been signed off, because they could not know which would be ready to deploy first.
>
> Instead of devising a system that would allow content to be present but disabled, they went with a hacked system (which I heard never worked properly) that was based on duplicate content branches and that required manual copying to resolve conflicts when the content was merged back together.

## Content management tools

The fifth level in the hierarchy of author experience needs focuses on content management tools. This level starts resolving questions of workflow, as discussed in the section titled "Workflow that… doesn't," and provides functionality to connect the dots.

## Solve approvals offline, first

The whole approvals workflow question is a nonstarter within current content management systems. Vendors claim they have solved workflow because they have implemented complex (or was that just convoluted?) interfaces that allow system managers to specify sign-off steps for content.

The challenge is not in implementing such an approvals process within a digital environment. It is that these processes are rarely formalized within the business environment prior to being coded into the CMS.

Approvals workflow is not a technical issue; it is a business-social-engineering problem. Some people think there is a magical digital process that will handle any content flow. But the people who need to interact with the content in such a flow are not computers. People are unpredictable. They will create situations the rules cannot cope with; they will bend and break the system.

What we need first, with regard to the approvals process, is to understand what we are trying to do and figure out a workable model that can exist outside the digital confine.

I do not offer a solution here: business process flow is beyond the scope of this book and, as you can imagine, is particularly susceptible to the specifics of each business. I suggest that it is really about risk management. The ideas espoused by Max Johns[23] on the subject – derived from the military, where lives are on the line – make a good starting point.

## Iterative content association

When I think workflow, I think of how information flows through a system. This is not the approvals process; it is the sequence that takes input and converts it to something suitable for output.

Imagine a product. You present information about it in your literature. How did the information get from the product specification to the marketing material? I am not referring to who told whom or who copied what. I mean the underlying process those people followed.

The initial technical specification is transformed into a field-based specification list, the sort of feature list that is used to compare similar products. This list is ordered and grouped. Then, for different output

---

[23] Max (@MxDEJ on Twitter) has posted a series of articles on risk management, which can be found at http://contentistheweb.com/riskmanagement/.

environments, parts of it are shown and parts hidden; terminology is adapted; and perhaps units are converted to suit the market audience.

The information flows through a lifecycle of conversions to adapt it for various audiences.

All of these transformations must be automated or at least linked. By linked, I mean that when an engineer discovers an issue with the product, any change to the master specification will flow to all the other systems that use that specification, through whatever transformations are relevant. (This process might not be entirely automated; controls may be needed to determine when the information is released to each audience.)

The information lifecycle is not about a block of information being created, published, and eventually retired. Information lifecycle relates to the transformation of information between audience versions. It requires keeping the references in place so that the relationship between source and derivative instances can be followed.

Imagine that you make a mechanical widget, and one of the dimensions is incorrectly noted in the specification but is being built correctly. If you can track that information to all the places where it is presented and, if your logging is sufficient, to all the customers who have ordered it or may have not ordered it because of the incorrect dimension, you are empowered to proactively fix problems.

## *Content, not pages*

This next improvement to author experience is almost an accidental derivative of associative structured content.

When defining references between elements of information, we have three choices.

The worst choice is to reference one specific page rendering in one assumed primary output. This has two issues. First, the output channel reference will need to be unpicked for all other channels and, probably, for the primary channel once the presentation model has changed. Second, what if the content we want to reference does not have any form of standalone rendering option? What if it exists only as something that is imported into a larger piece?

While the lack of a default (standalone) rendering may not occur frequently, there is a model that works just as well for content that has a default rendering as for content that does not: reference content entities for their own sake rather than identifying them based on a primary, page-rendering context. This is the second option, and it must be the minimum level of referencing used.

Because some entities do not have a default rendering URI, using a contextual presentation model instead of referencing content using a URI or path may gain some traction. Again, it makes sense to use a model that works generally and is more understandable than a model that might be easier to develop but caters to only 90% of situations.

> **You can't have that alone**
>
> In my sites, I have a Person content type. A reference to an individual imports a sidebar block with personal details, links to social media, etc. The Person may be reused in several articles. But there is no stand-alone rendering of the Person content type.

The pinnacle is to use dynamic associations, defining relationships based on metadata and rules without specifying an exact entity to reference. This third option has all the benefits of referencing content entities rather than primary renderings and provides additional flexibility (see the section titled "Content association paradigms").

## *Multi-axis content filtering*

To reference content instances, rather than pages, we need to implement dynamic multi-axis content filtering, not only for navigating the information store, but also within linking interfaces. Because we completely divorce content management from any output format that was previously considered primary, all interactions with the information store must be abstract and output agnostic.

I've heard arguments that multi-axis content filtering is hard to implement compared with the select-from-tree-list model, which is a default component in many programming languages. I've also heard that custom implementation would have more bugs and require more code.

These arguments may be valid, but they hold little weight if the easier implementation only allows us to link to one element at a time. And if you have ever had to use a tiny tree-model window that must be scrolled horizontally or that has so many elements at the same level that you can no longer see the parent hierarchy, you know it is not fit for purpose.

> **The pain of selecting from a tree view**
>
> While the window interface for a tree-select mechanism is bad enough, there is a further complication, which I have dealt with for several clients.
>
> When opening the tree selection dialog, should we:
> - Open the tree to the last reference that was selected?
> - Open the tree to the current selected location for the reference we are editing?
> - Open to the root of the tree every time?
>
> As you might imagine, there is no easy answer. Each of the three options appears to be optimal in about a third of cases and a serious pain in the neck the rest of the time.

Dynamic multi-axis filtering may provide an interface that is more convoluted than a tree view, but it provides so many benefits that it cannot be passed over simply because it is more difficult to implement. (And once implemented, it is just as reusable as a tree-view selector.)

Perhaps the most important detail is what an author can do with a metadata-based reference selector that could never be achieved with a tree-based reference: dynamic associations.

While it is easy to think of a link reference as a pointer to one specific instance of content, there is no reason the link reference can't be stored as the set of search and filter criteria that provides the final selection list. If it contains no results, one result, or multiple results is irrelevant: we have display rules for each case.

## *Flexible triggers*

A byproduct of the current crop of CMSes, and their propensity to include specific links, is a lack of responsiveness to the larger content ecosystem.

When first created, a metadata-based reference may resolve to a null pointer, referencing nothing. But when we create content that associates with that pre-existing reference, it will then resolve to a non-null, useful reference. When references and links evolve to be based on metadata, rather than always being specific, our ability to communicate will evolve for the better.

In principle, there is nothing fancy about this; it is not revolutionary. The concept that all references should be metadata-based associations, and that null-pointer references are valid, is merely evolutionary.

The next step is far more interesting. I am thinking here of algorithmic associations: links based on conditional rules. An algorithmic association could be as simple as a subject affinity combined with a requirement that the publication date be within the last week; the time-frame requirement means that links automatically become invalid after the defined period. Or it could be something more complicated that considers the subject affinities that a particular end user has followed, matched against other users' behaviors, to determine which affinities from the current content are likely to be most relevant to that particular user.

What value does this bring? It is not only about reference linking between elements but about functionality within the authoring environment. Dynamic, rules-based references would be invaluable in the governance of large information sets. They would allow author-relevance considerations to be added to multi-axis content filtering.

It is easy enough to envisage rules based on content age and last revision. Imagine how much more powerful this would become if we were to include analytics in the equation. Not only tracking how much various elements of content have been accessed, but tracking user interactions based on them (e.g. sales conversions) and feedback from multivariate testing.

This kind of complicated algorithm is nothing new. This sort of thing can be built into a CMS now. But an easily configurable version, in an out-of-the-box framework, would add significantly to the functionality of content management.

## Self-aware content

The pinnacle of author experience is a system that appears to understand the content that is managed within it and, therefore, picks up much of the management work peripheral to creating the message itself. To really improve life for authors, we need a structure that makes our content appear to be self aware.

## *Dependency awareness*

The most important feature needed to ensure that the system picks up some of the workload of managing content is dependency awareness.[24]

Most systems support only the most simplistic form of dependency awareness: if a piece of content is published and it references a secondary asset (image, video, etc.), then the referenced material is also published. If a CMS vendor claims to support dependency awareness, this is likely all they mean.

To understand what I mean by the term, consider a picture in your image library that is referenced in several articles. As those articles get deleted, the image is used less and less. Eventually, the last referencing article is removed from the system. What should happen to the image?

There is no simple answer. It depends on your governance.

The important thing is that the image know it is no longer used so that it can ask the question.[25]

The process triggered may be manual or based on some rule. What matters is that the event of becoming orphaned is recognized and can trigger a process. The details of the process itself are less important.

> A CMS must help clean up the mess resulting from managing content.

While media assets are the most obvious form of content to which dependency awareness algorithms could be applied, they are not alone. The same idea can be applied to other content types, with a twist.

When a piece of content is orphaned – no longer accessible via any navigation – how should it be treated? Again, the answer depends on many factors: is it intended as standalone campaign content? Is it linked to from elsewhere (externally)? Does it link back to the rest of the content? What was its purpose anyway?

To make this more complicated, we have a more abstract situation, where a group of content elements that reference each other are orphaned as

---

[24] Indeed, it is the lack of dependency awareness in a certain market-leading web CMS that first woke me to the entire field.

[25] True, it is the system rather than the content that will figure this out, but from the author's perspective, what difference does it make?

a cluster. This type of dependency separation also needs to be picked up because the cluster needs to be treated appropriately.

You might be thinking that this is getting painfully complex. I appreciate that. The alternative is a CMmess.

Putting complexity aside, I am going to take dependency awareness a step further. So far, we have looked at the idea of content knowing when it is no longer referenced. Realistically, those references have generally been considered to be direct references. (I didn't say they were, but it is a safe bet you thought of them that way.) The truly complex form is trans-algorithmic dependency awareness: content that looks backwards through filtering and selection algorithms to determine if it could ever be referenced by a filter or algorithm.[26]

Once we get to this type of dependency awareness, we cannot rely on author actions alone to trigger a dependency check: what if the algorithm involves a time offset, such as publication within the last 30 days? Yes, there is a lot of calculation that needs to go on for this. But we are talking about computers here; it is their job to do the computational heavy lifting. Let them get on with it.

## *Contextual content grouping*

Previously, I have mentioned situations where content is logically grouped: a set of elements that belong together; a set of elements that might need to be previewed with another, similar set; or a cluster of content that could become orphaned as a group while retaining references within itself.

This idea of a content cluster is missing from any existing CMS I am aware of.[27] But clustering is central to how content developers think.[28] Information does not become relevant one piece at a time – at least not from the perspective of those publishing it. When information is published, a set of related elements passes from the authoring environment to the publishing environment together. Whether or not they will be

---

[26] I am not aware of any CMS that includes good dependency awareness, let alone trans-algorithmic dependency awareness. Of course, I would be happy to learn otherwise.

[27] Most of the technology to achieve it is in place, it just hasn't been tied together to enable the functionality.

[28] One of the most obvious cases of clustering is a new product that consists of both direct product information and support material. Content clustering could be the key to building bridges between content ownership silos.

accessed in such sets by the end user is secondary. The cluster is defined by a shared context.

The obvious approach to creating a content cluster is through metadata. But I'm not suggesting an implementation model; that will depend on the CMSes internal paradigms.

The important thing is for the CMS to serve the authors' needs efficiently, to enable them to identify and see all elements that are relevant to a contextual group, to preview the release of a certain class of information, to run through whatever form of approval workflow makes sense within the system, and to publish it all. For that, they need a straightforward way of working with such clusters.

## *Learning systems*

So far, I have not suggested anything in terms of author experience functionality that is particularly arcane. None of the elements I have identified would be much of a stretch from an everyday programming perspective.[29] This next suggestion takes the complexity a large step further.

We need content management systems that keep track of what authors are doing, how they are doing things, which functionality they use, and which they don't – on a per-author basis – and then adapt the interfaces to provide each user with optimized functionality. I am talking about systems that adapt their interfaces to the user – that may show different subsets of the control interface to the same user based on different content types because they have learned what this user is most likely to do.[30]

Does this malleable interface concept present issues? Of course it does. We need to build a CMS that can track, analyze, and determine the basis for such functional adaptation. And we need to determine how to deal with consistency when we are being intentionally inconsistent.

I will not go into great detail about what I see a learning interface accomplishing. Realistically, I think successful implementation of such an evolution in computing is several years off, and once it starts to appear,

---

[29] Meaning that I could write them myself, perhaps not overly efficiently, but the basic mechanisms are straightforward enough. (The only reason I have not is that I really do not enjoy coding.)

[30] Around 2003, Microsoft experimented with this concept, though in a generic rather than learned way. The menus in their applications were shortened to only display the popular commands, requiring an expand feature to access the rest. (It was not popular.) Done properly, items would be hidden as a result of an active lack of use.

it will still need more time to mature.[31] All I know is that we are talking about personalized adaptive design, sufficiently aware of the author's behaviors that it optimizes in every situation.

## *Helpful suggestions*

In parallel with the evolution of personalized adaptive interfaces, I believe author experience will also be enhanced by systems that can understand and predict the meaning of the material they manage and can identify themes and concepts that human authors may not be aware of.

What does this mean? Imagine that you have written the description of a product and referred to a particular part (say, the case). Now, there is nothing special or unusual about this part. It is so bland that it does not need to be described. Everyone knows what it is. But over time, you write about a few more products and refer to the same part.

The part (case) is so uninteresting to you that as an author, you see no value in tagging it with metadata. Not the helpful CMS, though. It spots the contextual repetition and even if you have not asked for it, it creates an internal reference that ties all of these elements together[32] against a future possibility of associative relevance. And every time you refer to a case, it is automatically, and invisibly, added to this reference set.

Then, one day, you introduce colorful replacement cases.

In the old world, you would need to hunt down all the affected products, find all the case references, and determine which ones need to be updated with a cross-reference to your new replacement cases.

With a CMS that understands your content, this process would be considerably faster. Your new cases need to be referenced in place of all case references for products that meet a certain set of criteria: one flexible trigger applied to an element of system-generated metadata, and the job is done!

Helpful suggestions made by the CMS are not limited to creating metadata. Smart mechanisms, when combined with learning systems, could provide an author with reduced lists of the most relevant and pertinent options, even if those options are not ones this author has

---

[31] Come back and ask me sometime around 2022, and maybe the industry will have evolved far enough that a sensible conversation can be had.

[32] This is a basic form of *latent semantic indexing*.

recently used. If other authors start using a new option, there is a good chance authors who haven't used that option yet will soon find it relevant.

The range of helping functionality that could be integrated into CMSes is constrained only by imagination and the willingness to invest in development. For example, imagine a system that can flag style-guide grammar issues.[33] Or one that tells you when you create duplicate content. Maybe one that can detect changes in staff and flag content responsibility for potential reassignment.

If we have a good implementation of iterative content association, then any legal elements (small print) would be cross-referenced to the source legislation, maintained in an unrelated system. When the legislation changes and the disclaimer requirements evolve, the CMS can notify the relevant parties.

Why not provide the ability to represent the full information set using relationship maps, based on associations the system detects rather than author-defined metadata? This could help find redundancies and gaps.

The possibilities are endless.

---

[33] Acrolinx [http://acrolinx.com/] provides this type of capability today.

# CHAPTER 6
# *Conducting an Author Experience Audit*

The purpose of an author experience audit is to objectively determine the adequacy of the procedures and tools used for content authoring to ensure the quality and integrity of the managed content. If we compare it with the other audit type common in the field of content strategy – the *content audit* – there are some distinct differences.

A content audit objectively analyzes information about your content: what it is, where it is, and how it relates. It maps relationships, determines subject coverage, identifies gaps, and analyzes content usage patterns. The content audit maps attributes of elements of content against the rest of the content set. The AX audit deals with the more subjective concepts of business requirements, author perceptions, and paradigms. The quantitative part of an AX audit compares the current state against an ideal design.

The author experience audit consists of five steps:

- Technical business analysis
- Content design and governance
- Author experience design
- Information architecture
- Auditing

Except for the author experience design step, the process uses established business processes. There are, however, dependencies between the steps: content analysis without business analysis may give you great answers, but they will lack foundation. Also, the initial analysis will be more valuable if undertaken with an awareness of the author-experience details that will emerge later.

In addition to the five steps of the audit itself, this chapter covers:

- Goals of the AX audit
- Tools of the trade
- Measuring author experience quality

## Goals of the AX audit

There are two types of author experience audit, but the processes involved in both are similar, and there is a natural flow between them.

- Audit of an existing system
- Audit of new system requirements

## Common aspects of AX audits

An AX audit considers communication needs within a business's domain, now and in the future, and establishes how those needs map to content structure and governance. It enables you to define consistent, business-specific paradigms that support managing your content.

You can then weigh the optimal content management model against either what is currently in place or the cost of implementing the optimal model. Based on costs and value, you can decide whether to fix or implement and when to do so.

## Existing-system AX audit

With the existing-system audit, you need to document two sides: the optimal model and the features and performance of the existing system. By considering how well these two lists match up, you can find excess functionality and gaps in the system. Of course, the definitions of excesses and gaps are not black and white; they depend on measuring the quality of the existing system's author experience.

In theory, you should remove excess functionality, though it can be hard to make the business case for this. This is especially difficult when content management is handled at the lowest rung of responsibility within an organization. The flip side to this is that while cleaning up the interface may make things more comfortable for those managing the content, the real value is in avoiding distractions,[1] thereby reducing the risk of poor or incorrect information being entered into the system.

Any identified gaps provide the baseline for new system functionality, evolving this into the second class of audit, the new-system audit.

## New-system AX audit

Auditing for a new system is a case of following a business analysis process: identify requirements and then transform them into models of the content management processes that need to be supported. From there, you can identify the mechanics of the functionality in the author environment – the interfaces, not how it works under the hood.

You can then cross-reference these functional specifications against your CMS or, better yet, use them to help select a CMS. And you can weigh

---

[1] You may think that people are good at ignoring irrelevant mess, and they may think so too. But if this were really the case, there would not be UX research showing the increased efficiency of streamlined, uncluttered interfaces.

the cost of each feature against the value that feature will bring to the integrity of your communications process.

The aim is to identify the investment required to make the system fit for purpose and weigh this against the content risks that would result from deploying a substandard system.

## Tools of the trade

This is the point where I have just one word for you: sorry. I cannot furnish you with spreadsheets to perform your audit. There is no single model that does the trick.

An AX audit is business analysis, combined with a solid understanding of technology (both technology in general and the systems that make up your stack).

In some instances one person can handle the whole process. But if the task is too large, you need a team. If you use teams, I recommend dividing the analysis into different end-to-end processes rather than dividing it by function (analysis, design, etc.). This is because information gained in earlier parts of the process affects later steps and vice versa. However, where this is not possible, it is reasonable to divide the job into the four main steps – analysis, content design, AX design, and information architecture, with the final auditing then returning to the business analyst – provided someone who understands the end-to-end process oversees it.

Many organizations consider that the experience of the lowly grunts tasked with managing content does not count as a business driver; those people are so poorly paid that their efficiency is irrelevant. Perhaps the most important advice I can give you for performing an AX audit and presenting the results is to demonstrate value in terms of the integrity of the information and the ability to communicate efficiently, and to downplay the value of improving the lives of the authors.

Of course, if you are dealing with an organization that already understands the value of content and, therefore, has people with greater responsibility (and pay) handing content management and authoring, the value of their time could enter the picture. But even then, the quality argument must still be dominant.

The real tools of the trade are your brain and your adaptability.

## Technical business analysis

Technical business analysis is a term used to describe business analysis relating to software systems. As with any business analysis, it involves asking questions, exploring the underlying reasons for requests, understanding the desired outcomes, and then devising solutions. It requires you to understand the basics of how systems work and the limitations or obstacles the environment may impose on your recommendations.

### *Why the information?*

The process of analyzing content management interfaces and processes starts far from the technical implementation. It starts with the content that is to be managed and the underlying business reason for having that content. We are not asking, "What is this content supposed to achieve?" We are instead asking, "Why do we even need content?" (And then, "Why do we need this content? Why do we need that content?")

Sometimes, these questions are easy to answer. Other times, not so easy. If your business is to provide news services – your content is your product – the answer is obvious. If you make widgets used in a process that has a potential client base of three companies (your entire audience is three purchasing managers), and you are creating a system for marketing material, the reason for doing so is harder to discern.

Answering this basic question is important, and most people can answer it. However, the approach many take to answering this question has two shortcomings.

First, the answers are incomplete, often driven by the wrong principles and concepts, because they come from asking about the content or the platform. Instead, they need to come from a model of the business's domain and where it intersects with the domain models of those who will consume the content. This is the most accurate way to understand what the information is, how it flows, and why it is needed.

Second, many people, once they know the answer, accept it and get on with the rest of their analysis. Instead, they need to do two things with this initial answer:

- Challenge it, to understand the underlying reasoning
- Keep it front and center throughout the subsequent processes, to validate the answers and decisions that follow

## Who, what, and where?

Once we understand why we need the information, we need to understand the audiences, the actual information to be exchanged, and the media and channels that will carry it. Note that information is exchanged, not presented; some elements need to be captured.

At this point, we need to map out the process flows around the information – how elements interact with each other, with other systems in our organization, and with external sources. (If we were producing only a single element of standalone content, with no relationship to anything else, we would not need to build a system.) We are not yet modeling the content types the system will manage. But we do need to understand what those types are and their relationships.

A news article relates to other news and likely to non-news content. It needs to be aggregated; it needs to be accessible via navigation; it is returned in search; it needs to reference an author and/or editor; and it may exist as part of a larger sequence of stories, a short update supported by back-story highlights.

We also need to gather some initial knowledge related to how our information is presented across media. This moves us towards content modeling. Our information informs the range of elements we can expect to need. For example, it can help us determine whether elements need different versions to cope with the limitations of different display environments.

Keep in mind that information is not only exchanged between the system and people. Other platforms are part of your content ecosystem. Is your information linking through to all parties who need it? Does information about a product sale propagate to stock control and manufacturing on one hand and to customer relation management on the other?

Sometimes, this integration will be manual, if it exists at all. Indeed, far too often platforms are implemented in isolation, with references at best copied and pasted between screens in different interfaces. Because this lack of integration incurs risks and inefficiencies, you must identify the changes required to other systems (which could include replacing them) to allow automated integration and determine whether the risk in making, or not making, those changes is acceptable.

Our analysis of how the information needs to flow and be shared with other systems may reveal the need for integration work that is not presently justifiable but will be needed in the long term. Then, we need to ensure our system is set up to allow those integration points to be easily added later, so we avoid having to reengineer both systems to fulfill the need. This may mean including data elements the eventual integration will require, even if those elements are not relevant to the non-integrated system.

And we must ensure that all these answers align with the underlying purpose of the system. Does the information we are planning to exchange fulfill a real business need, or is it the result of an ego trip?

## The wider value

It is easy to see, when we realize that all integration points need to be mapped and considered, that information is the lifeblood of any enterprise; it flows up and down chains and weaves its way between silos and departments.

And as the whole point of any content management system – or really, any digital system – is to facilitate the manipulation and movement of information, we see that whatever our system does, in an optimal world it will reach into every corner of an organization, managing all the informational relationships.

And you thought you could just deal with a few words on a page!

# Content design and governance

The second stage of the AX audit is to identify the types of content you will need to manage within your system and create models for them. This may sound straightforward; occasionally it is. But structured content is often significantly more complex than people care to realize.

Am I saying you must make your content models complex? To some extent I am, but not without caveat. Information, and information management processes, are inherently complex. After all, they are carried out by decently complex beings: people.

Model your information's full complexity first and then determine where you can simplify. This is better than aiming to devise a minimum viable model from the beginning[2] only to discover later that you missed key references and structures.

---

[2] We still want to end up with a minimum viable model; the difference is that, because we are dealing with fundamentals, we want to start with everything and work down rather than building up until we think we have enough. Yes, this goes against some current trends. With good cause.

Consider a news article; what elements does it consist of? Here are some obvious wrong answers:

- A title
- An author name
- A body

These elements constitute the core of a news article, yet I have identified these answers as obviously wrong. Why? Because we need to consider these things in far greater detail to model our article content type.

Is there only one title? How do we present our title within the article (be that in print or online)? How much space do we have? How do we cope when the title is longer than the space we have available? Do we have truncation rules, or do we identify more than one title with different length guidelines?

How do we display the title in a reference to the article?

What if our article is part of a larger story? Do we include the definition of the larger story into the main title, or do we have a separate field for that? (In a support documentation scenario, this is the distinction between "solving problem X for product Y" and "solving problem X" when the scope of product Y is already inferred by how and where the content is presented.)

Dealing with the author is a little easier… at first. We do not want an author identified by name; we need a reference to the author. The reference then provides access to the information we want to present: the author's name, biography, photo, and a list of other articles.

To make matters worse, we need to consider the reality of news-article writing. How many authors does an article have? Always one, but sometimes more. Is the author a staff reporter or a news-gathering organization? Are all authors equal or are some only contributors? And how does this evolve if the article has a long shelf life, getting updated by different writers and editors over time?

This is an example of the distinction between the true complexity of the information and how much of it we need to make available. If we decide to handle only some of these scenarios – based on how often articles have multiple primary authors, have contributing authors, or get updated by others who take over responsibility – then the other aspects of the model can be discarded. But at least they are highlighted, documented, ascertained, and removed from scope.

With company publications – thought-leadership content – we run into other problems around the author. When an individual leaves, does the company still want the ex-employee's name associated with its material? Likewise, does the individual still want his or her name associated with an ex-employer?

Lastly, we come to the body, but calling it the body is a gross over-simplification. As we saw in the section titled "Sequence and narrative flow," we need to define the structure of the body. We need to know what elements are allowed within it. Are there restrictions – sequences that are not permitted or mandatory elements? Will we enforce those rules through programmatic restrictions or through author training?[3]

Once we have identified our content types, their constituent elements, and the references between them, we need to consider the business processes involved in managing all of these things through their lifecycles. It is not enough to say that we can create content. Do we also need an approvals process? Who is able to publish the material? Who can reuse it or edit it? How do we deal with content when it is less relevant, and what do we do when it is totally defunct? When can we completely delete content, or must we keep reference versions for legal or regulatory purposes?

Defining the governance model for a piece of content might inspire changes to its field structure. With the news article, we have a publication date (an obvious field). But what happens when we correct an error? Do we republish with a new date or continue to rely on the original date? And does that answer depend on how serious the error was? Do we want to indicate that there was a correction with an updated-on date?

Basically, governance processes are an integral part of the content model. You cannot say you have designed the content structure if you have not considered all the interactions of the content with processes, and with other content, that could occur throughout its lifecycle.

> Content that does not have a governance model has not been completely defined.

---

[3] While it is nice to think that programmatic restrictions are a perfect solution, it is easy to miss scenarios where they inhibit information.

## Author experience design

Only when we know the scope of the content our system must manage and everything that content needs to do and have done to it can we look at the structural paradigms that best suit the authoring process. Having all the information from the technical business analysis and content design and governance steps does not guarantee you will get the AX right. The reason – experience tells me – is that content governance is always underestimated;[4] few realize just how much work it involves, so they make unrealistic promises about what they can do regarding generating and maintaining content.

> **The governance workload**
>
> I could rattle off multiple examples of clients who have made promises (to themselves) to maintain published information. They almost all fail to live up to those commitments.
>
> But instead of giving examples, I refer you to the majority of the internet, replete as it is with cobweb sites.

### *Reused paradigms*

When designing the author experience, two of the first and most obvious subelements of the interface to consider are: how authors will find content and how they will create links and associations between elements.

It is impossible to give a single solution to these issues. Really, it depends.[5] The most important factor is the quantity of content you have to manage. If you have a single block of content, you need only one link. Expand this to a handful (no more than you can identify in a quickly scannable list) and still not much is required, because the mind finds the right entity faster in a short list than it would take to enter keywords in a form and search for it. The list can extend to a small tree, if the distinctions at the first level are sufficient that there is no room for confusion (people, products, events – only).

Once you have more content, you need some form of dynamic list with search/filter properties. Add more content with more axes to filter on, and the filter mechanism itself becomes abstracted, enabling the author to select which filter attributes to employ.[6] At this stage, it probably

---

[4] And content longevity is always overestimated.

[5] Please excuse the swearing. There is no other way of telling you that truth.

[6] Just because there is a primary means of finding content does not mean that you cannot surface secondary mechanisms. If, for example, you have a preview function that allows authors to navigate your content, there is nothing wrong with providing links from that

becomes relevant to include a précis in the results list, because content names will not necessarily provide enough information to distinguish between elements.

The same approach applies to linking content together. The linking interface should resemble the interface used for finding content to edit. After all, both activities are fundamentally about finding content.

An interesting twist here is that sometimes the CMS exposes authors to different linking paradigms for types of content that should be treated the same. For example, you may need to reference images or other multimedia assets differently from the way you reference other elements of content. This is usually the result of different storage models within the CMS. As discussed in the section titled "The CMSes dirty underwear," the storage models should be irrelevant; we don't want to expose internal paradigms to authors.

The flip side of this twist is that sometimes the CMS uses the same paradigm for two types of content that should be presented to authors in different ways. I am thinking here of a web CMS that stores metadata tags as nodes in the same way it stores pages. Because the concept of the tag-attribute is different from that of a page, these two data types should not be selected or managed in the same way.

You also need to consider consistent interfaces for error or confirmation messages and embedded help. How does the system show that content has been published? That it has been edited? How does the author know that there is a published version but the latest edits are not complete or that there are unpopulated mandatory fields? Each of these communications between the system and the author needs to be considered and designed. While it is nice for them to be intuitive, it is more important that they be consistent and coherent.

## *Sub-structured content*

The next phase of AX definition is to consider the full structure of each content type and how best to approach authoring each of them. But it is also about the structure within the structure. If your content type has

---

to the editing interface. Likewise, existing reference links between content entities should provide authors with the ability to open the target entities for editing. When dealing with secondary access routes, bear in mind that they will probably not be available for linking. If they are generally more efficient than the primary, you create a confused paradigm. Whatever access route authors prefer should be optimized and used as the main content search mechanism.

a hundred fields, how are these arranged for authoring? What kinds of groupings are used?[7]

Again, this is about consistency.

Sometimes, you may find that one content type has only a small number of fields, but these map to a more generic model where each is part of a different sub-structure. Do you include the hub-and-spoke model that works for the larger sets, but with each spoke containing only a field or two, or do you surface all the fields in a single interface?

There is no single answer. The goal is to ensure that the paradigms authors learn when editing one content type carry over seamlessly to other types. If the system is upgraded and a new content type is added, the author's reaction should be "How did I not notice this before?" rather than "How does this one work?"

These questions of structural consistency apply not only to the basic create and edit interfaces but also to the governance controls within the content instances and to any content listings.

## *How much detail?*

If you adhere rigidly to the arguments I have made throughout this book and go for the hyper-granular approach to content structure, you could convince yourself that you need hundreds of fields within your content type structure to maximize its efficiency.

I am sure that this would be too much detail for most authors to manage. The whole point of author experience is to facilitate the authoring process, making authors more productive and less prone to errors. An endless list of fields does not accomplish this.

A strong argument in favor of reducing information sets comes from a social engineering experiment carried out at Cook Country Hospital in Chicago in 1996.[8] Cook County Hospital is the place of last resort for hundreds of thousands of Chicagoans without health insurance. The lack of insurance brings many patients to the Emergency Department, and a significant number of these patients believe they have suffered a heart attack.

---

[7] In the aside titled "Counting Titles," this was a modal dialog within the larger parent content type.

[8] The tale and learning related to Cook County Hospital are paraphrased from *Blink: the Power of Thinking Without Thinking*, [Back Bay Books, 2005], by Malcolm Gladwell.

Faced with a life-or-death choice, doctors tried considering all pertinent factors when deciding whether to admit patients to the coronary care ward or send them home. With so many known heart-attack risk factors, the limited beds were more often than not filled with patients who needed no further intervention.

The experiment tested a model devised years earlier by cardiologist Lee Goldman. Goldman worked with a group of mathematicians to analyze data relating to chest-pain complaints and correlated that data with the actual outcomes. The results – and the system implemented at Cook County in response to those results – simplified the diagnosis of whether a patient had a significant risk of having a heart attack within the next three days. Most factors applied to the long-term risk, not the immediate probability, of a heart attack. Goldman found only four factors that had any real influence in the short term: an abnormal ECG, the type of pain, fluid in the lungs, and low systolic blood pressure. From those factors, a consistent decision could be made that was significantly more accurate than the opinions of admitting physicians, despite their years of experience.

What does this have to do with content management?

Simply that the number of fields we include in our structured content types might be overkill, especially with regard to metadata. We run a real risk of overwhelming ourselves with so much metadata that the increased likelihood of errors overshadows any benefit we might gain.[9]

The amount of metadata to include in your content is obviously a difficult question – one with repercussions. The more content you have to manage, the more metadata you need to differentiate that content, which means additional fields or a wider range of available values. There is no one-size-fits-all answer. We need the minimum viable distinction between content elements and no more.[10]

And just as it took Goldman a lot of research and analysis to find the key factors to identify the risk of a heart attack in the short term, so you will need to do a lot of analysis to understand the most meaningful, relevant, and manageable ways to differentiate your content.

---

[9] Think of it as the 80/20 rule applied to the relevance of metadata.

[10] To get to the minimum viable distinction, we are better off starting with too much granularity and paring back rather than starting with too little and building up to that minimum.

## A definition of mandatory

I recently heard a presentation by Todd Copeland, Senior Director at Experis Global Content Solutions,[11] where he provided synonyms for mandatory and optional fields: *mandatory* is "a burden to the person putting the information in," and *optional* is "ignored."

Over the years, I have seen many instances of content types with large numbers of mandatory fields – some had dozens – and they prove Copeland right. It is unreasonable to expect authors to populate so many mandatory fields in one go. I've seen cases where it took several sessions, spanning days, to complete an element of content.

Yet the rule – because it was the easiest to program – was that all mandatory fields had to be populated before the content instance could be saved, even when the content was in an interim state. Sometimes the rules were a little looser; only individual substructures needed to validate to save progressive changes, but again, this was an unnatural requirement.

The consequence of this populate-to-save approach is that authors use filler: information that certainly must not be published. They then waste time reviewing and re-reviewing their input to ensure that every last instance of filler has been replaced by something that is correct.

The correct application of mandatory fields is easy to devise:

> Mandatory means that the field must be populated before the author moves the content to the next phase in the governance lifecycle.

The existence of an empty mandatory field should not stop content from being edited and saved; it should only restrict content from moving to the next phase in the governance cycle. This immediately reduces the risk of inappropriate filler content ever being published, because there is never cause for filler to be entered.

## To form or not to form?

All this talk about structured authoring raises an important question: what will these interfaces look like? With so many fields, are we facing death by forms?

National Public Radio's COPE (Create Once, Publish Everywhere) is often held up as the poster child of how to do reusable structured authoring right. But looking at a screenshot of a COPE interface (see Figure 6.1),

---

[11] https://www.brighttalk.com/webcast/9273/107957

you might have reservations. The stacked input boxes give it the feel of an old-style database: raw and unforgiving.

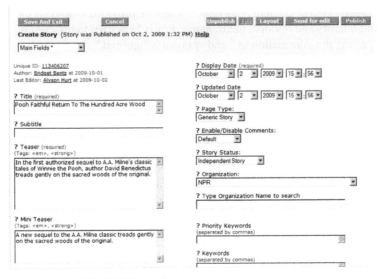

Figure 6.1 – NPR COPE interface

With so many fields required for a single content type, we must have forms. There is no other approach. But we can make forms work better.

I believe the trick is to present our forms so they do not feel like forms. We want the same structure and functionality without the appearance – and psychological impact – of raw stricture.

There are two aspects to this.

The first element is the visual presentation of the form itself. When I say form, you may be thinking of a label and a rectangular box. But what if we were to soften that? What if we were to have a blank sheet with some headings and placeholder text (so authors know where to start editing) and fields that expanded to accommodate what we added to them?

This approach is being used by GatherContent[12] to make their content aggregation tool author friendly (see Figure 6.2). It results in a process that feels more like building a story than filling in a form. The mechanics

---

[12] https://gathercontent.com/

are the same, but the perception is not. And when it comes to ease of use, perception is significantly more important than the steps involved.[13]

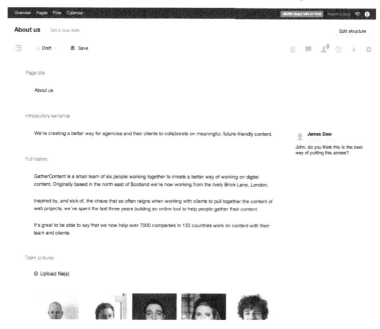

Figure 6.2 – GatherContent screenshot

The second detail that differentiates easy-to-use from cumbersome forms is how much is exposed in one go. If you split a form into subsections, each of which contains only a few interrelated fields, and display only one subsection at a time, the result is less overwhelming than one long interface.[14] Appropriate clustering also reinforces the context of the elements being exposed.

And if you want authors to use your CMS anywhere other than at a computer with a full sized screen, you almost certainly need to offer some type of hub-and-spoke authoring paradigm. Using a single-field form on a mobile phone is hard enough when the keyboard takes up half your available space, and let's not talk about those selectors where you scroll through 100 options.

---

[13] Margot Bloomstein's *Content Strategy for Slow Experiences* talk at CS Forum 2013 showed how frustration is a more accurate indicator of perceived duration than is time. http://slideshare.net/mbloomstein/content-strategy-for-slow-expreiences-cs-forum-2013

[14] As with anything to do with changing perceptions, this is very easy to overdo. Too many too-small sections becomes just as problematic.

## *Integrating analytics*

A subject I have largely left aside so far is analytics. Analytics measure the value and performance of your content: how end users interact with it and how effective it is in achieving its aims.

There are many content analytics tools available today, especially within the web domain. In the days of print and broadcast media, the effectiveness of a message was determined by the aggregated response of one's audience; in the modern world, businesses want much finer detail – often to the point of customizing for each individual using a complex profile garnered from a variety of sources.

I won't go into the details of how analytics work or how to use them; this depends on your content and the performance indicators you have established. But I do want to cover the key points that relate to analytics as they apply to digital platform delivery and author experience.

There are two ways of looking at analytics. The first approach considers displayed pages as the fundamental element, whether these represent individual pieces of content or aggregated blocks. It lets you see what pages have been viewed and how successful those pages have been. This is a top-down view of usage. The second approach considers the granular elements of content – cross-referenced with the pages they were displayed on – to analyze their success. This is a bottom-up view of usage.

If your content is managed on a page level, the two approaches are equivalent. But with content reuse, the second approach offers insights that cannot be garnered from the first.

The bottom-up approach to analytics requires integration between your CMS and the analytics platform. If you only need to view the performance of a piece of content within the context of managing it, the CMS needs to query the analytics platform. This is generally straightforward.

To use the analytics platform's functionality – its dashboards and comparison tools – to perform bottom-up analytics, you must ensure that your analytics provider can establish a secure connection with your CMS. This connection enables the analytics platform to obtain human-readable content identifiers.[15]

---

[15] You might think that you could simply add human-readable metadata to each granular element of content and use that to record usage. With a large content store, this requires more and more unique human-readable identifiers, which become difficult to manage (not to mention being an additional mandatory field). It also risks exposing information about your content and processes to the public or your competitors.

## System capabilities

The last step in designing the author experience is the one I often wish were not a factor. But until those needing new information systems do the analysis up front and make their platform purchasing decisions based on needs rather than ticking boxes for snazzy features, we need to deal with this issue. Sometimes, the platform we are required to work with simply can not support the concepts we – as author experience professionals – consider optimal.

We can work around some of these out-of-the-box limitations through system customization. Others are simply too fundamental to the platform to be overcome, and we need to adapt our AX paradigms.

The most important point here is to keep the adaptation consistent.

I opened this section by implying that the reason we may need to make compromises in our optimal AX is that the CMS is often selected before the content management requirements are defined. This is not entirely true. Even if we have a full specification available to drive the vendor selection process, it is unlikely that any system can do everything we want without customization. Sometimes, it will be more pragmatic to adapt our requirements than to insist on all the customization work needed to get things exactly as we envisage them.

So, until a CMS can abstract content sufficiently that any AX-centric paradigm can be configured within it, compromises will be demanded. And made.

## Information architecture

It might seem appropriate to stop the author-experience audit process at the AX design stage. But that only tells us part of the story. This next step formalizes the earlier elements and provides the documentation against which existing interfaces can be judged or from which new systems can be built.

The term information architecture (IA) has picked up some rather poor connotations over the last decade: there is a perception that it is about creating wireframes for web sites and little else. That perception is like saying the plasterer builds a house for you. It omits the need to lay a foundation, build the frame and walls, install plumbing and electricity, and hang doors and windows. Then it leaves out painting, decorating, and landscaping.

Information architecture is not only about the one interface required to present your information to an end consumer. It is the architecture of the actual information itself – the logical associations, paradigms, and interactions in the bowels of the storage and manipulation systems that handle it from input to output.

As this book is not about information architecture – there are enough treatises on that subject elsewhere – I cover only what the IA part of the audit is intended to deliver.

Through the AX design process, we were focused on how the logical elements of information we are managing are associated – how an author perceives those relationships to meaningfully manage them. But our content needs to exist in two other places:

- It needs to be presentable to the end user.
- It needs to be stored within the system.

The information architect reviews the content design and considers how to model it for storage and manipulation within the CMS and also how to present it optimally. Then, the information architect defines the transformational mapping between these models (see Figure 6.3).

> Information Architecture covers the structural form of information within an environment and the translation to its interfaces.

It is easy to think that there is one best way to store and manipulate information. But there are huge differences in logical implementation between a database-centric content design and an XML-in-structured-tree design. The granularity of the elements used, how they relate for optimal reuse, and many other factors come into play here.

The amount of information architecture work needed depends on whether you are auditing an existing system or performing a pre-build audit. With an existing system, it depends on whether the current interfaces come close to delivering a meaningful AX or if they need to be heavily reworked.

Importantly, unlike the way you handle the author-focused aspects of the process, to do the IA work, you need to identify the platform. You need to know what the underlying technology is: how it works, what it can do, and what it cannot.

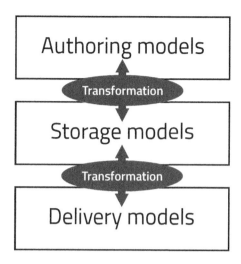

Figure 6.3 – Information architecture: translating information structures between environments

If you define the author experience before selecting a platform, then when you write the RFP, you need to include the authoring models and paradigms that the platform needs to transform its data storage structure to and from.

## Auditing

So far, everything we have discussed has been preparation work for the audit, establishing the baseline of what our system must do.

The audit considers the paradigms and structures defined within the analysis, content design, author-experience design, and information architecture. From these, it determines the extent to which the current system matches those needs, or it establishes the implementation path required to build such a system.

Assuming that an exact match between the baseline and the existing system does not occur, the audit must determine the effort required to align the two. For a new system, it establishes all the elements needed. Unless you have never dealt with any kind of CMS implementation, you know that the requirements and available resources rarely match, which means you will need to prioritize.

To prioritize, you establish which types of content require the most management and, in particular, what content you will reuse (i.e. reference) most. You must also establish dependencies accurately, as the effort to retrofit a dependency can be devastating. Of course, there are always counter-examples: if a system is sufficiently complicated, it may need to be built incrementally, without all the foundations in place first. Sometimes, functionality-enhancing dependencies must be retrofitted.

> **Prioritizing dependencies**
>
> Some years ago, I was involved in the development of a customer relationship management system for a headhunting firm. This was not just recruitment. It was board-level headhunting; clients and candidates were drawn from the same pool.
>
> When I was brought on board and shown the development plan, I spotted a rather nasty issue. Someone had decided that the security system was to be the last thing implemented (or even designed!). Despite warnings, no one bothered to change the plan.
>
> As you can imagine, when it came time to implement this most fundamental piece of the platform, everything needed reworking to accommodate their security needs.

When you complete this process, you end up with a prioritized backlog of functional elements to implement.

## Measuring author experience quality

To successfully audit author experience, we need ways of measuring what we observe. As with anything labelled experience, there is no absolute objective measure of the quality of author experience. All such measures are inherently subjective and relative. Individuals have their own perspectives on what is important and what is the relative value of different improvements. But there should at least be general consensus that one experience is an improvement on another.

### *Paradigms*

Because a large part of the challenge discussed in Chapter 4 has to do with how authors think about and envisage the content they are working with, our first measure of author experience quality relates to how much sense the system's surfaced paradigms make to the authors.

This can be judged using two simple criteria:

- How much training do authors need to start using the system?
- How long can authors be away from the system without forgetting how to use it?

To ensure that our CMS really supports authors in their business processes, we must ensure that it is easy and intuitive to use. We must surface obvious paradigms rather than internal system logic. The system must resonate with the business's *domain model*.

You may think this is overkill – that your authors are specialists, trained to use the environment. They have technical skills. And while this may be the case in some organizations, for some types of content, this model cannot survive. Just as the business executive's personal secretary has become a rarity, so too does the guardianship of digital content systems slip out of the hands of dedicated content managers. Everyone interacts with the content system at some level. And so the system needs to make sense to everyone.

## *The fit-for-purpose system*

Another way to measure the value of a content management system is to consider how well it serves its purpose. Does it facilitate content management?

Much of the answer to this is subjective. For the individual whose mindset is sufficiently different from that of the rest of the team, the otherwise wonderful system may feel illogical. However, two criteria can objectively measure how well the system fits its role:

- To what extent does a single, simple content management environment support multiple output channels?
- How often must authors manage or follow associations between content elements manually (as opposed to the relationship being displayed automatically)? Being able to manage associated elements without realizing they are distinct is even better.

## *Value of an improvement*

When a change is made to the author interface of a CMS, there are several ways of measuring the value.

We could consider the time savings authors experience in performing their task within the system. Unfortunately, this comparison sets itself up to be weighed against the cost of implementing the change. Given the cost of development and the incremental nature of the savings that result, this is not a particularly favorable measure.

The second measure is the authors' perception of how usable the system is after the change. This presents challenges:

- This measure can be evaluated only after implementation.
- Many organizations are reluctant to spend money on staff experience and satisfaction when more immediate results could come from spending on customers.
- The change from old to new takes some getting used to, so the improvement may initially prove disruptive.

A more persuasive argument of the value of an AX improvement is direct return on investment: evidence (or the suggestion) that the enhanced system will increase revenue or cut secondary costs (such as translation).

Perhaps the best measure of the value of an AX enhancement – the one most likely to work with those controlling the purse strings – is how well it helps the system fulfill its purpose. Does it reduce the risk or need for content duplication? Does it eliminate repetition? Does it reduce the possibility of inconsistencies and the resultant corruption of your message? Does it reduce the losses from bad content?

## *Upgradability of the system*

Another measure of author experience quality is whether the underlying system can be upgraded without reworking all of the AX customization. With minor upgrades to the underlying system, this is rarely an issue, but major releases often come with significant changes to the way content is stored and manipulated.

The vendor usually covers the changes resulting from a version upgrade if – and only if – the system has not been customized. Part of the version upgrade process is the remapping of content and management paradigms from old to new. But this still leaves any author experience enhancements as the client's responsibility. The threat of future cost coerces many clients into adopting the vendors' content management paradigms – a behavior the vendors encourage because they know that the upgrade costs could easily be redirected to moving instead to a different platform.

The counter to this is to ensure that author experience – and the customization of the interface to suit the business's needs – is brought up early in the negotiations. If your vendor guarantees a consistent AX-to-system translation layer or agrees to own the burden of ensuring portability of the customizations, then you will be able to upgrade the underlying system without significant issues.

> **No upgrades here**
>
> I have been on the periphery of several discussions about system upgrades. In each case, the client wanted the new features (usually performance and security), but the cost of performing the upgrade was prohibitive: it required several weeks of testing and, usually, redevelopment of significant amounts of the system customization.
>
> Upgrade costs equivalent to the original build costs were not inconceivable. (It probably didn't help that the developers were slower the second time around, because they needed to learn how the new version worked.)

## *Upgradability of the business*

The second way to look at upgrades with regard to author experience is in the evolution of business processes. No business sticks with one set of processes for ever. Whether your CMS is customized or out-of-the-box, the author experience must not lock your business in to a single chain of thought or process.

Your communication needs will change. If your author experience restricts your ability to change how you communicate, if it limits the business's ability to upgrade itself, then it fails.

# CHAPTER 7
# *Author Experience Design Patterns*

Design patterns provide general, reusable solutions to commonly occurring problems. They are conceptual approaches, rather than finished products. Design patterns originated in architecture and are common in software development. We can apply the principle of design patterns to almost any field.

In this section, I outline various design patterns that I, and others, have used to improve the author experience of certain content management systems. Not all are relevant to all scenarios, nor is the list exhaustive. I have included nine patterns here, which fall into three categories:

Micro-copy management and usage patterns

- UI label management
- Integrated label use
- Label structure deployment

Author interface patterns

- Content consoles
- Narrative
- WYSISMUC
- Selection by date

Technical patterns (as far as they impact the author)

- Referrer - referee links
- Stored or searched references

## UI label management

The question of how to manage user interface (UI) labels – also known as micro-copy – is relevant to almost every project. There are two aspects to this: managing labels for the end-user interface and managing labels for the author interface. Here, I am dealing primarily with the former.

This design pattern is related to fit-for-purpose language, associative structured content, and content management tools in Chapter 5.

## The challenge

Micro-copy may sound trivial; it is anything but. The labels used in your content delivery system's interface provide context for your end-users, helping them understand, navigate, and use the system.

Often, we give too little consideration to the definition of these labels. In many instances, they are hard coded into the templates used to serve content to the end user. Hard-coded micro-copy is problematic for three reasons:

- Applying changes across the system is cumbersome.
- Localization is difficult.
- Editing can only be done by a developer.[1]

That micro-copy must be managed as content is obvious. We have a potentially very long list of labels; we need a convenient, logical, and flexible way to manage them. Key to this is the ability to distinguish between labels and understand what each is used for.

## Why it matters

Micro-copy sets the tone for the presentation of information. It expresses functional and navigational meaning. The difference between buttons labelled *send* and *submit* may sound trivial, but end users can infer a lot from this, which could affect their perception of your organization, even to the extent of determining whether they will do business with you.

Micro-copy is generally reused throughout an information system. Multiple buttons with the same functionality use the same label. As such, a submit button label is not for a specific form; it is for a class of forms. But micro-copy is just a list of labels maintained as content, without context; we cannot rely on a label's values alone to explain its function or correct value.

This is made more difficult when translating. The lack of context can be devastating: any single word or group of words can be translated several ways into most languages. Lack of context leads to errors. Also, the same label value may be used for different elements in your source language, but different labels may be necessary in other languages.

---

[1] Or at least someone with developer tools, and in my book, anyone with the tools, and permission to use them, is a developer, regardless of job title.

## *An approach that works*

To manage micro-copy, you need a dedicated content type.

This type needs some form of categorization, based on functional or thematic use of the labels. For example, you may have some micro-copy used for support and some used for an online store. Of course, things get messy when you have elements that span categories or categorizations that span existing areas (for example, navigation micro-copy could be used for both support and an online store). How you deal with the categorization and separation depends on your situation and how much micro-copy you have.

The micro-copy content type is special. Unlike other content, which authors create as needed, micro-copy is created by developers. Micro-copy is relevant only when it is used by a template, so it is created only when something references it. Deleting micro-copy is also a developer function because you do not remove a label from your content store until no templates use it.

The interesting part of this content type is not the creation or deletion role, but that there are two editing roles for it. One role is to author the micro-copy text. The second role is to provide the default value used when the label is not populated and to edit the contextual descriptor for the label. The contextual descriptor is metadata that explains the context in which the label is used.

The key aspect of this pattern is that the contextual descriptor is used as an authoring UI label for the content itself, to clearly identify the field being edited (see Figure 7.1).

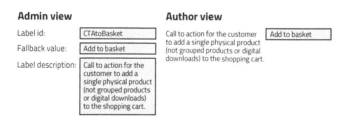

Figure 7.1 – Contextual descriptor used as an authoring label

For example, if we have an "Add to basket" button in our online store, the contextual descriptor for this field might be something along the lines of: "Call to action for the customer to add a single physical product (not grouped products or digital downloads) to the shopping cart." As a general rule, the contextual descriptor is significantly wordier than the value of the label it identifies. It describes the function associated with the micro-copy and the purpose of that function.

As a minor aside, while I have identified three roles in managing micro-copy, most organizations can safely combine the developer's create-delete role with the administrative edit role. Descriptions are modified rarely enough that you usually don't need a separate role. Developers can be asked to take this role, or you can train a few authors to handle it.

### Authoring labels

You should only rarely need to edit author UI labels (Label id, Fall-back value, and Label description fields in Figure 7.1), so these fields should not, generally, be editable. The only reason for making these labels editable is that micro-copy is developer-created content. We cannot assume that the final contextual description of all labels will be available when the labels are first identified.

If possible, separate the definition of the author UI labels from the functional code used to manage other content.[2] It does not matter whether the format of this separation is developer-friendly rather than author-friendly; once a content type is released to authors, the author UI labels rarely change. However, when they do change, it is far easier to manage them if they are not embedded in code. Also, if at some point in the future you might consider localizing your author environment, then having the author UI labels separate from the functional code will be a boon.

## Integrated label use

This design pattern is related to content accessibility and rules-based presentation as discussed in Chapter 5.

### The challenge

The technical documentation for your product refers to the user interface labels when describing functionality. Suppose, for whatever reason, that a label needs to be changed. The system and its documentation cannot be changed independently without causing confusion.

---

[2] However, it is generally not worth developing the functionality for a single deployment if it does not already exist within the platform

## Why it matters

Consider a product with a visual interface and a set of documentation that describes that interface. Both depend on the micro-copy. This raises the question: who owns the micro-copy?

Sometimes, it is those responsible for the product. Sometimes, it is those responsible for the documentation. Or it could be a third group. We need a solution that works regardless of which party has authority. And we need a solution that reflects changes in both the product and the documentation automatically.

## An approach that works

We have already established that a CMS is really any system that stores manageable information. As such, we need the labels to be managed within a CMS and reused within both the product and the documentation. It doesn't matter what CMS you use or whether you manage labels as part of the product build, the documentation system, or a third environment. What is important is the integration.

And that is the point; this approach requires both automated integration and support for label references in all systems. Manage the labels, and always use references to import the correct values.

Unfortunately, this is only part of the solution.

Your label management environment will probably need to support different label values for different versions of your product, and the integration processes need to know which version of the labels to use for each product version.

Additionally, if your documentation includes screenshots, you need a second level of integration. The images must be updated to reflect the current label values. The key factors are how much and how often your micro-copy changes. That determines whether updates need to be handled manually or automatically. Even if the process is manual, screenshots must include sufficient metadata to ensure that when a label changes, the system identifies all images that must be replaced.

## Label structure deployment

In an enterprise environment with multiple layers of development, test, and production, we will need to move labels between various platforms. This design pattern is related to content management tools as discussed in Chapter 5.

### The challenge

In an enterprise setting, content management systems are convoluted. These systems usually have three or more layers – versions of the platform where functionality is developed, tested, and put into production. This layering is separate from the author/publish split we have considered so far, which is itself present within each layer.

For most of this pattern, we consider only two main environments: the development environment and the (production) author environment. The other distinctions are of little consequence for this design pattern. This pattern derives from web content management systems, which usually have a fully functional author environment. However, the implementation is applicable to all systems.

When we are dealing with enterprise content management, especially with ongoing functional and content development, different mechanisms are used to move functionality and content between environments. Functionality is promoted from the development environment to both the author environment and the production publish environments by copying code and, perhaps, by manipulating existing content storage structures. Content generally exists only in the author environment, and publication is a simple copy to the publish environment.

As we saw above in the section titled "UI label management," label content differs from most content. We must manage two separate aspects – the definition of the label and the management of its value – within different environments. This can create a mess.

### Why it matters

As we saw earlier, managing labels is a three-part process. First, you define the labels and create fallback values. Then, you define the contextual descriptor metadata. And finally, you populate the label value that will be presented to the end user.

While this process sounds straightforward, it has a catch. The three functions occur in different environments.

Developers define and create labels in the development environment. The definitions are then deployed to the author environment.

Where contextual descriptor metadata is managed varies, depending on the organization. Some organizations consider metadata to be structure and manage it in the development environment. Others manage metadata in the author environment. (I will not deal with the contextual descriptor specifically in this design pattern, as it copies the behavior of one of the other parts of the process.)

The label values are managed in the author environment, as any other content is handled. Therefore, different environments have authority over the label definition and its value.

We cannot simply copy content between environments. If we change the definition of a label in the development environment we need a way to change that label in the author environment without overwriting the label values managed in the author environment. We also need a way of moving the label values defined in the author environment back to the development environment, against the normal flow of deployment, without affecting any changes to the label definitions that may have occurred within the development environment.

## *An approach that works*

UI labels are content. But part of that content is a structural definition, managed in a layer that generally does not see much real content.

To deploy UI label content from the development environment to the author environment, that content needs to be treated as structure. But we cannot simply copy labels as we would other functionality or structure. We need the system to actively manage the process.

A newly defined label is simply copied in its entirety, and deleting a label is equally straight forward. But when a label already exists in the author environment, only the metadata is updated; the label's value must never be overwritten. This is because the author environment is the authority for the label value; the value can only propagate outwards from there.

The propagation of the value leads us to the second part of this pattern.

The UI label values managed in the author environment need to be available in the testing and development environments.[3] We cannot use

---

[3] When developers are testing functionality, it is often useful for them to have real content – especially micro-copy – to test against.

the mechanism used to publish UI label content for this transfer. The publishing mechanism copies the entire UI label content item because the publish environments need all the metadata. Again, we need to process the content when we deploy it from the author environment to the development environment. The processing rules in this case are simple: copy only the UI label value. Do not copy any metadata. And if a UI label from production does not exist in development (it has been deleted), do not recreate it.

This allows us to manage the parts of the UI label content type in different environments and propagate them between those environments without creating conflicts (see Figure 7.2 for a simplified illustration showing just the author, development, and publish environments).

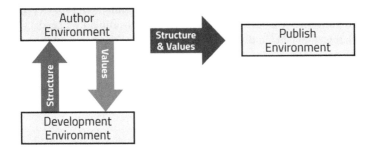

Figure 7.2 – Deploying label structure and value

## Content consoles

How a CMS enables authors to navigate is a subject I covered in the section titled "Content accessibility." Where the content set is small, it does not matter much – a dozen content entities can be more easily considered and selected from a single, flat list than they can be found through search. The real strength of this next pattern comes when we are considering thousands of pieces of content.

This is also a meta-pattern: a toolbox of elements that work together in different combinations based on the specifics of the platform and content types being managed.

This design pattern is related to the section titled "Content accessibility" and the section titled "Content management tools" in Chapter 5.

## Author Experience Design Patterns   125

### The challenge
Too much content. There is no way to stay on top of it all, to even know what exists, or to know what you want to edit, reuse, or reference. With so much content, we need to be able to find things we didn't even know we were looking for, using searches that are logical, natural, and do not enforce a particular categorization.

We want to find content that matches what we are thinking.

### Why it matters
When content is managed by a large team, there are many unknowns.

It is not always possible to keep track of every element of content that is being created. We cannot know what everyone else is doing, and we may not know whether the content we want exists. And that means we have a high risk of duplication and omission. We need a well-structured approach to categorization and retrieval that does not rely on understanding how someone else may have defined the content.

We need to find what we want quickly and effortlessly every time. And if we can't find it, we need to be confident that what we are looking for does not exist.[4]

### Approaches
As I pointed out previously, the optimal way of finding content within a large store is through some form of multi-axis filtering process.

This method leaves us with the key question of what axes to filter on. Some of these may be obvious and universal, while others depend on the content type and other factors.

Obvious axes involve such attributes as free-text search, content type, creation date, publication date, publication status, authors, editors, current owner, and so forth. Not all of these are relevant for every project, and not all of them are relevant for every content type.

---

[4] There is still the risk that two people will look for the same thing at the same time, not find it, and both create it. I am not going to offer a solution to concurrent creation conflicts here; while it may sound like an important case, it occurs so rarely that it is not worth the effort to implement at this scale.

## The partial approach

We also have context-specific axes such as categorization attributes, references to or from other content, review scheduling, and field-specific value matching. This provides us with a lot of search options.

The down-side is that the list of suitable axes will be different for each content type. If only we knew the type, we could identify the dominant axes and a reasonable range of optional axes.

A partial implementation of this pattern, which is often good enough, involves identifying the content type and defining, for each type, a set of filtering axes. The obvious catch is that if we do not know the type of the content we are looking for, it becomes much harder to find.

## The best solution approach

Identification of a content type is relevant only when we create a piece of content. After that, we think of it as a nebulous entity that happens to behave in certain ways. We are more interested in the specifics of what it is about than what type of structure it is. As such, the requirement to identify the type first when looking for content is counter-productive.

There are counter-arguments to that. Some content types, for example a person and an article, are simple to distinguish. But what if we need to distinguish between an article and a press release when searching for content? The difference is small enough to potentially be confusing.

So how do we deal with this?

A system that can do this will require a lot of development, which is probably why no one has yet built a particularly good model. As outlined in the section titled "Dynamic multi-axis content filtering," we need a model that accepts a search value and then provides contextualization options to refine the search based on the available matching results. This requires well-structured content and active management of the contexts implied by metadata – a task that is only justifiable if you are dealing with massive quantities of content.

You may be thinking that we could simply settle for a Google-style search. The problem is that we would need to manage synonyms, which would take as much ongoing investment as managing contexts. So, until computers can understand meaning and find the content that matches the ideas we are trying to express, we may be left with only partial solutions.

Perhaps more important than the optimized filtering is the ability for authors to save search parameters. We can then extend the pattern to make it more useful to authors; that is, enable them to save their personal filter patterns and even set their own defaults per content type.

## Narrative[5]

I have mentioned a few times earlier that a key aspect of communication is narrative, building a flow into the message. This design pattern is related to associative structured content as discussed in Chapter 5.

### *The challenge*

The ideas of narrative and structured content may appear to be at odds with each other. Structure puts up walls and determines boxes we need to fill in, whereas narrative is a good story, a flow that adapts to the needs of the message being conveyed.

How can we get these to align?

### *Why it matters*

This is as much a stylistic issue as anything. We need our narratives to have consistent structure if they are to fit together. Just as a good story sets a scene, introduces characters, builds a challenge, and evolves towards a resolution, the narrative structure of each piece of content within a larger repository must follow patterns that impart information in a certain way or evoke a particular emotional response.

Whether we are providing advice on design patterns or trying to sell something, our communication has structural models that identify both the type of content and the organization behind it. These must be balanced with the specific demands of each piece of content – the inner structure that best communicates its particular message.

### *An approach that works*

We need to balance two things here: the need for a tight structure and flexibility. We can accomplish this easily by looking at the structure of a sample of our content type. What sequence makes up the narrative?

By sequence, I do not mean paragraphs, images, and tables. I am asking the question from a conceptual level: introduction, background, core

---

[5] This pattern is the result of considerable input from Elizabeth McGuane (@emcguane on Twitter), combined with my own experience.

information, advanced information. The answer to this question divides the narrative of our content type into zones.

The elements of content that can be used within each zone now need to be determined. We can display information in various ways: paragraph text, lists, images, interactive widgets, tables, etc. We end up with more questions. Which ones belong in which zones? How many elements can each zone contain? Are there limits on the order of these elements?

By defining rules for elements and zones (which, how many, sequence restrictions), we achieve flexibility within structure.

The key to this is to provide choice, but not too much. A little choice empowers, too much confuses. Except for text-only zones, most zones should allow some flexibility – a choice of two or three options for representing information. Also, if there are dynamic widgets available within your structure, these should be easy to configure (think along the lines of a multi-image slideshow: images, captions, done).

## *The complicated part*

Depending on the locale where your content is used, this model may need to support some interesting quirks. In some cases, the rule will be that the localized version must have the same content as the original, just directly translated. But sometimes – especially with marketing content that talks to the social context of the audience – you need to localize even the core of the message. You may have a situation where you simply cannot translate the content because it won't work for a local audience.[6]

In this instance, we find that different parts of the content may have different localizations treatments. Some localizations will retain the source structure, while others will completely replace the original content with audience-appropriate material.

Unless you want to give your development team headaches, I strongly suggest making zones the finest granularity on which to distinguish which source must be retained and which can be replaced. Do not try to lock some elements within a zone but not others.

---

[6] Your motorbike is marketed in the USA as a recreational vehicle for twenty-something men who want to push social norms, so performance statistics and a lifestyle video are relevant. But in India, a motorbike is an efficient and effective means of commuting for everyone, so alternate content is needed.

# WYSISMUC

This pattern deals with a means of authoring and managing content that bridges the gap between the presentational/visual forms that people are inherently comfortable with and the semantically structured authoring that a digital platform needs to make content reusable. WYSISMUC stands for What You See Is Structurally Marked-Up Content.[7]

This design pattern is related to associative structured content and content management tools as discussed in Chapter 5.

## The challenge

As we have seen before, people generally think in terms of the whole message, including presentation as part of the content. As such, they feel most comfortable creating content in a visual form that bears some resemblance to how they expect that content to be displayed.

When the only output was print,[8] it made sense to use WYSIWYG. The problem with WYSIWYG is that the information is stored in a way that references the intended medium's presentation. As editing evolved from print to the web, the WYSIWYG paradigms persisted: the storage may have evolved to use HTML rather than a print instruction format such as Postscript, but editing remained coupled to a presentational model.

> To make content portable, it must be stored within a declarative semantic structure, devoid of presentational instructions.

## Why it matters

Content is fluid. It needs to adapt to myriad devices and platforms. If content is structured for one particular output model, even if that is the most common, it must be deconstructed and then reconstructed whenever it is to be delivered using another output model.

You might be thinking that there is little difference between taking a semantic structure and converting that to each required delivery format and deconstructing a primary format for the conversion. And indeed, that could be the case when the primary format is highly structured and follows strict rules.

---

[7] In some circles, this is referred to as WYSIWYM (What You See Is What You Mean). And you will find other acronyms that say the same thing: visual editing that is semantic.

[8] True, there was never a time when the only output was print, but there was a time, not so long ago, when content would be created for one medium at a time, with no desire or expectation that it would be reused.

The problem is that primary formats rarely follow structural rules. Your content has semantic relationships that the format simply cannot cleanly represent.[9] A presentation-focused editor generates content markup that lacks semantics. It gets messy. Anyone who has had to clean up the XML or HTML generated from Microsoft Word (especially older versions; they are getting better) knows what I am talking about.

We need editor interfaces that are somewhat presentational in their display, but capture authoring actions in semantic ways. The interface needs to enforce context-specific semantics.

## An approach that works

The solution is WYSISMUC. This is an evolution of WYSIWYG that will look similar to authors. The main difference is the markup it generates around the content.

The first change is in the functions that the editor makes available. In WYSIWYG, there are many formatting options: bold, italic, underline, super- and subscript, fonts, font sizes, colors, alignments. Additionally, you can create lists (bulleted or numbered), indent or outdent content (indenting creates either a blockquote or nested list, depending on what is indented), add links, identify header levels and paragraphs, and insert items such as tables and images (with identifiers as to size and position).

WYSISMUC replaces most of those controls with ones that are more precise about the structural meaning of the content. For example, bold, italic, and underline will be replaced by emphasis types (it is *what* they do, not *how* they do it); the author must indicate the semantic function of the emphasis. Super- and subscript might remain, but only in specific instances where they are warranted. Font, color, and alignment controls, including heading types, are replaced by semantic heading descriptors. We keep the paragraph, but the headings themselves are given meaningful names rather than arbitrary numbers.

And when headings are used, the editor manages the logical block it identifies automatically: it knows which of the following paragraphs belong to the block defined by that heading.

---

[9] If you think that HTML is a good format for your primary content, please tell me how you meaningfully attach a footnote to the content. Remember that while the footnote is displayed at the foot of the content page, its logical position within content flow is hanging off the side of what is being read. It needs to refer to the location to which it is attached.

Things get more interesting with lists, quotes, images, and tables.

When creating a list, it is not a case of saying, "I want a bulleted list," the list itself has a semantic type, which must be identified. Is it a list of ingredients? A list of steps in a process? A list of people? Some lists would then allow sublists and some would not.

Quotes, images, tables, and any other sub-elements have two forms. They can be elements within the flow of the content, which belong between two paragraphs, or they can be figures attached to particular paragraphs or sections (as defined by the blocks that headings relate to) – effectively metadata.

Footnotes and links become metadata on a selection of words within the body of a paragraph. In effect, we end up with mixed media content that is associative rather than purely sequential.

No single WYSISMUC editor suits everyone. There is not even a single editor that is appropriate for all instances within a particular content store. The editor is highly configurable. Which media elements are permitted? What list types are available? What types of semantic emphasis are relevant? Are links just one reference or may they be series of references? Must an image be accompanied by a caption?

The result of using a WYSISMUC editor is semantically structured content, generated without exposing authors to code.[10] WYSISMUC provides a sense of narrative while keeping elements associated with each other in a meaningful way.

The editor makes no demands as to how the content is stored within the CMS. But there is a requirement that the stored format be converted to and from the editor's internal format (which, if it is appropriately stored, should not be an issue; the editor format is just another rendering). How each of these elements is eventually rendered is dependent on presentation rules, per medium.

---

[10] It is not only that there is no need to expose authors to the code. We also must ensure that there is no access to the underlying content-code generated by the editor. The ability to override the structure is an invitation to make the content non-semantic.

## Selection by date

Dates are a straightforward way to categorize information. They can tell us when something was created, edited, and published. But, what does a date really tell us?

### *The challenge*

Computers manage and manipulate dates as a number – an offset counted in milliseconds. This is not conducive to human understanding, so we have them translated into more meaningful formats like days, months, and years.[11] Unfortunately, once the date is translated into a human-readable form, we lose a key piece of information.

If I pick a date – yesterday, for example – what period does this cover? It is *obviously* the 24 hours from midnight to midnight. We're done.

Except we're not: a detail is lacking that would make the obvious answer correct, but without which a more accurate answer is 50 hours.

### *Why the confusion?*

Most people live in an insular temporal bubble.[12] They know their local time, but have little concept of other time zones. Without knowing which part of the world the yesterday we are referring to is in, we do not know which 24 hours to consider, so must consider them all.

There is some overlap at the International Date Line. Kiritimati, on Line Islands Time, is the first place to greet a new day; its time zone is UTC+14.[13] Baker Island is the last place to see out each day; its time zone is UTC-12. Baker Island begins a new day two hours after Kiritimati has started the next day. Therefore, the full range of time that a particular date could refer to is 24 hours plus the sum of the most extreme offsets: 24 + 14 + 12 = 50.

---

[11] An example of one place where author experience has been reasonably implemented for about as long as the public has been working with computers.

[12] If you do not believe this, just consider how daylight saving/summer time is handled now. Most clocks adjust themselves. Time zone identifiers are losing their daylight/summer differentiators (US TV channels now simply refer to ET, rather than EST and EDT; Europeans have for years used CET – Central European Time – year-round, even though seven months of the year are really CEST). Google Calendar invitations include a comment about not being adjusted for summer time, which, if read carefully, says the opposite of what is intended.

[13] UTC (Universal Coordinated Time) is close enough to GMT (Greenwich Mean Time) to be used interchangeably for everyday consideration.

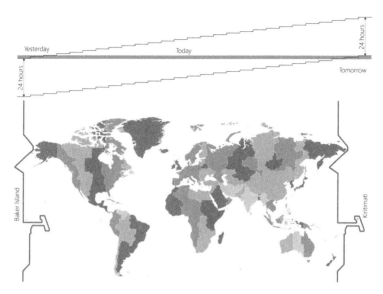

Figure 7.3 – Which day is it?

In Figure 7.3, we see that it is yesterday, today and tomorrow… all at the same time.

## Why it matters
What does this have to do with content management?

As humans, we think in terms of days. We rarely qualify our referenced date with the time zone it applies to. This is not an issue until the people dealing with your content are spread around the world. Then, a reference to the day something was created, edited, or published becomes fuzzy.

If you search for content by date, what time zone will it match on? Where you are? The arbitrary time zone your server is configured for? Or will it cover all of your content team's contexts?

## Approaches that work
There are two basic approaches to this issue.

The first is to present the time stamp on any content in the user's local time zone. Unfortunately, sometimes people learn what date to search on through a source in another time zone. A person in Australia tells someone in the USA the date to search on, but they are 15 hours apart.

The second approach is for the system to understand this problem and expand its searches to the 50 hours that any given date could imply. For this, assuming the server is standardized on UTC, any given date extends from 10:00UTC the previous day until 12:00UTC the following day. This window can be shortened if your authors do not operate at the outer reaches of time-zone offsets.

## Referrer / referee links

This pattern relates to a simple aspect of author interface design: links and references. This design pattern is related to content accessibility, associative structured content, and content management tools as discussed in Chapter 5.

### The challenge

When you are managing a piece of content, how do you know what other content it refers to – whether directly or through filters? Or what content refers to it? (Because the interface pattern is the same, I focus on outbound references primarily and look at inbound references only at the end.)

In many cases, to create a reference you must open an edit dialog and find a reference link that is presented, as likely as not, as a URI path.[14]

### Why it matters

References and links are key to usable information. While a single piece of content may be self-contained, connectivity and interaction generate the most value.

And just as it is important when consuming content to have those links, it is equally important that the authors managing that content know what links and references are in place without having to decipher some code or truncated URI.

They may also need to access the referenced content, at least to review it and possibly to edit it. Accessing referenced content is typically a detour – an aside in a larger process.

---

[14] To make things worse, it is not at all uncommon for the field displaying the reference URI to be too small, showing only part of the structure: the root path elements that do not distinguish one from the next.

## An approach that works

Within a content entity, there are several ways to reference other content. From an article, there is the metadata section, which includes authors and cross-references; there are separate elements that identify any images the article uses; and the article body may include links.

The simplest approach involves remembering the purpose of author experience: to facilitate content management. This means that we provide meaningful information that is understandable to the human mind, not the technical underbelly of the implementation. Show a snippet from the referenced content rather than the reference and always include a link to open the referenced content for editing. Do this both within any edit dialog and also from the larger view of the whole content item.

Where the link is an abstract filter (such as you have with multi-axis content filtering) rather than a specific target, the display should run the filter, and the link for editing the referenced content should not open just a single target. Instead, it should open a list that contains all the results the filter could return, from which the actual items to be edited can be selected.

Even if you are using a tree model to select content to link to, clearly distinguish between the view of the selected content and the selection mechanism. Once something is selected in the tree, show the selected content. (If you embed an image in a piece of content, you expect this behavior, so extend that model to all references.)

Figure 7.4 – A reference that does not provide meaning

Figure 7.5 – A reference that clearly identifies the content

There may be places where this kind of verbose reference cannot be rendered inline with the content. This is understandable. Sometimes, you need to compromise.

### Inbound references

Inbound references can be presented as a list of content references – think search results – which are only retrieved when requested by the author. Since this list could be fairly long, pagination or grouping may be needed.

## Stored or searched references

This last pattern is largely a technical implementation question, but one that can have a huge impact on authors.

This design pattern is related to content accessibility, content management tools, and self-aware content as discussed in Chapter 5.

### The challenge

There are many references between elements of content within a CMS. If something happens to a reference target, some action may be required on the source of the reference.

To automate such actions, or at least to notify an author that such action is required, we need dependency awareness. This raises the question of the best way to obtain a list of content that references a particular item.

### Why it matters

You can manage and, more importantly, store references between content items in several ways. One option is to create a database table of references that points to either end of the reference. To find all the referenced to a piece of content, all you need to do is query this table.

But many CMSes use other mechanisms to create references. They may include the reference as an embedded link, almost anywhere within the content structure. Then, the idea of a search is more difficult, as we do not have a single, optimized place to look.

Performance depends on the size of the content store and the underlying technology. If small, then a search is as good as instant, and no one is worried. But if your content store consists of thousands of content items, and every time a reference is changed a dependency awareness check is required, then the few seconds required to parse the entire content store searching for instances of the reference quickly add up, making the system cumbersome. Those few seconds of delay frustrate, and frustrated content managers are more likely to make costly mistakes.

## An approach that works

This solution may be controversial from a development perspective, because it goes against what many programmers consider the right way of doing things, but the objective here is getting the best performance in large systems when you don't have a database of references.

Where the system does not contain a database table that identifies all references and where the size of the content store is such that parsing it for references takes more than a tenth of a second,[15] then all references should be double stored. When a reference is created, data is added to the target to show that it is being referenced by the source. This way, it becomes easy to generate a list of every link coming into a given piece of content.

This solution starts to break down when the references are filters, though for some cases a filter may be a simple interim referrer that provides a layer of indirection. In that case, targets could keep a list of filters that will return a reference to them.

Where this solution breaks down completely is with algorithmic filters, which return different results depending on the user's context. On the other hand, in this situation everything can be referenced by just about anything, so tracking references becomes largely irrelevant. The best option at this point is probably a compromise: capture the references that are easy to track, and accept some uncertainty around the edges.

---

[15] This is not a scientific cut-off; it is about perceivable delay of the interface. I have suffered systems where it was closer to 20 seconds.

# The Future of Author Experience

# CHAPTER 8
# *Moving Forward with Author Experience*

Author experience as a specialized field is young. I know of only a handful of people paying real attention to it; a few more are involved in a peripheral way, mostly as an afterthought to content structuring, end-user UX, or content strategy.

It is a field that calls for different justifications for different stakeholders. Corporate leaders may not care about the happiness of lowly content editors, but they respond to the risk associated with unfit-for-purpose tools. Editors are (usually) interested in making their own lives easier.

It will take time for author experience to become a recognized and expected part of every content system project.

In this chapter, I cover various arguments and challenges we will face along the way. Some arise early on, as we are working towards general recognition of the need for this discipline. Others rear their heads any time we get a seat at the project table.

## A new name for old rope?

Previously, I mapped out the steps in an author experience audit. Technical business analysis, content design and governance, and information architecture are all reasonably well understood. And the author experience design step could, if one wanted to ignore a few key distinctions, be called user experience within the author environment.

But the combination of these four streams deserves a name of its own, especially when someone brings all four of these elements together in a unified process. There are savings to be had from performing the business analysis with a clear understanding of what the content modeling repercussions are likely to be and how they will affect the AX design. This understanding enables you to identify and mitigate conflicts between business needs and the platform's limitations earlier in the process.

In truth, it does not matter what title an individual holds when designing a solid and meaningful author experience. What matters is that whoever performs the business analysis understands how author experience affects the content design and how technical paradigms affect subsequent phases. In other words, this individual must approach the task as an author experience designer.

## The customers are more important

I have repeatedly stated the purpose of author experience: to ensure that information management interfaces are suitable to the people using them, to reduce the business risks associated with poor content.

Nevertheless, there will always be pressure from all levels within an organization – especially when faced with budget restrictions – saying that "the customer is king," so end-user experience is a higher priority. After all, how can the principle of user-centered design (UCD) be anything but gospel?

I happily admit that UCD is a powerful and relevant toolset, but it is an approach fraught with risk. UCD that is not counterbalanced by BCD will run away and create great experiences that do not further the organization's goals.

Hold up… what is BCD? Business-centered design, of course.

BCD provides the constraints within which UCD can effectively operate. It identifies the purpose of the system and provides business justification. When applied correctly, it ensures that a company keeps its communications and interactions with customers focused on areas that benefit the business, through direct commercial interactions or brand development.

When you bring business principles to the table, two things come out:
- It becomes harder to justify a lot of the fancy user-focused features that do not serve the business purpose of the system.
- The quality and accuracy of the information receive higher priority.

## It's too much work

Developing good author experience takes work, especially from coders. They are accustomed to surfacing the information model that makes sense to them. Adding a translation layer to their model to make it author friendly feels like unjustified effort.

In reality, the problem here is that most teams do not have a framework for converting between the underlying model and an author-friendly one. Getting such a system in place requires a massive effort, which is hard to justify for that first small instance where it will be used. And as the framework evolves to cope with more flexibility and options, the economies of scale take a while to be realized. But eventually, with a decent framework in place, the effort will shrink significantly.

If the developers continue to insist that it is too much effort to develop the framework, let's regress the environments they work with to the same level as the interfaces they think are suitable for content management and see how they like those tools.

## CMS vendors

The last of the preemptory challenges to getting author experience to be a successful discipline in its own right is the ongoing fight we are going to face with system vendors. In an industry largely dominated by the sales tactics of a handful of suppliers touting the latest buzzwords and extolling the features and approaches that work for their own content-modeling paradigms, the idea that content managers must have a say in the structure and approach to their own work is anathema.

We need vendors to understand that a translation layer is needed between their content paradigms and the authoring interface to allow easy customization of the author experience so that it fits its intended purpose. And this must be done without creating an impediment to upgrading. To accomplish this, I believe we need clients to go into their RFP processes with authoring paradigms defined and documented so that vendors are measured against real-world requirements.

If vendors can understand that their success or failure depends on their adaptation to business needs, rather than expecting each business to modify its processes to conform to the vendor's paradigms, we will eventually have systems designed to be easily customized for an optimal author experience.

And once one of the bigger vendors does so, the others must either follow suit or lose business.

## Conflict of author interest

You might think that the concept of author experience – of making life easier for those who must use the CMS – would be universally welcomed by the authors themselves. After all, who in their right mind would object to an improvement in working conditions?

Surprisingly, this is not always the case.

With one client, the primary author contact consistently pushed back against any improvement to the author experience. Any suggestion that might have automated repetitive tasks or reused content or references to avoid duplicate content was always rejected as unneeded.

I never did get to the bottom of this. It is hard for me to understand this mindset, but my best guess is that there were two possible reasons. The first had to do with job security; the second was based on a cost-benefit model that, at least to me, appeared skewed.

My instincts tell me that the rejection of better tools was about the authors' jobs. The team was not particularly burdened, so anything that removed risk, thereby automating some quality assurance and oversight, might have resulted in layoffs. Of course, their thinking could have been more about the perception of competence: their content management roles were so difficult, they couldn't be automated, which meant the staff themselves were "highly skilled professionals." And such skills need to be appropriately compensated.

The other possibility is that even though they were small, the proposed improvements were seen as costing significantly more than the people doing the work. And authors did not perceive the risk to content integrity as important enough to justify that cost.[1]

Whatever the true reasons, the principle remains that we cannot assume that authors will support the changes we propose. There are always factors at play that we cannot anticipate.

## Silos (Who owns content?)

Who owns content? Who is responsible for creating and maintaining it? More importantly, who is responsible for ensuring that content ownership is respected when different departments need to reuse each other's content? Who ensures that departments with overlapping – often conflicting – responsibilities and agendas play nicely together? If the department reusing content needs to adjust it in some way, is it now their content, or does responsibility stay with the original department?

These are not easy questions to answer.

Some organizations have created a new Digital department that owns the content management and digital channel publication technologies, adding a new head to the corporate hydra.[2] As a department in its own right, Digital becomes just another competing player with its own agenda.

---

[1] Even though I heard several tales of errors that required them to roll back content deployments, the team seemed happy to go through all that extra work.

[2] If you doubt this, just try searching for "Chief Digital Officer." At time of writing, Google returned over four million hits on the exact phrase.

In other organizations, each department claims responsibility for its business-critical information, purchasing or implement systems for its needs without considering overlap with others. There is a lack of coordination. The systems each department implements cannot communicate or share content.

A few businesses take a more mature approach: they see technology and the information in communication systems as the lubricant that enables everything to work (look no further than the likes of Amazon for an integrated information company), but they are exceptions.

Author experience comes at the questions of information responsibility from a holistic perspective. AX wants to be a support service for the whole business, ensuring that all the wheels are greased and information can flow into every corner to everyone's benefit. We are the ones arguing for the best solution, modeled on the whole business domain, while other players are interested only in the bit of the puzzle that suits their immediate needs. So our efforts are often undone by politics, and the value we bring to the table is lost.

## The Agile battlefield

The last challenge to implementing author experience relates to process. It is the result of the mass adoption of Agile as a system implementation methodology. This is not a Waterfall vs. Agile rant; it is a comment on the success of Agile and the resulting expansion of the range of activities managed with an Agile methodology.

As outlined in Chapter 6, the AX process requires analysis, design, and eventually paradigm specification. We need to know, before we dive head first into the implementation of our CMS customization project, just what we intend to deliver.[3]

This approach is aligned with the Agile definition: you must start with a full backlog of defined – buildable – stories to choose what to implement. But when Agile is put into practice, the principles that defined Agile can get muddied.

The first issue has to do with over-delivering on the base principles of Agile. One principle of the Agile Manifesto reads "Working software over Comprehensive documentation." But, in my experience, many

---

[3] This section considers the impact of using Agile to develop your author environment. If your content team uses an Agile methodology, you must build your workflows to support Agile, which is a different discussion entirely.

practitioners interpret that principle as saying "End-user demonstrable software…" rather than "Working software … " and focus on the ability to demonstrate each iteration to the non-technical client. This can lead them to omit the critical step of building a strong foundation. This is analogous to building a house by doing the paintwork first, before laying a foundation or even having walls to paint.

This is an understandable deviation from core principles; many clients struggle to understand what is being delivered when they cannot see the results.

The second is that Agile has become a victim of its own success. Since Agile methodologies have worked for development, the concept has been extended to Agile design. This says you can start with a high level concept and design its functionality in an Agile manner while building it. The idea is to start the development cycle sooner, thereby – allegedly – delivering the product faster.

This approach assumes that we can use testing and feedback for analysis and quickly correct functionality and interface design. While the first part of this assumption may be true, the second is not true within a CMS authoring environment. Instead of the functionality being reworked, it accumulates as *technical debt*. Also, failing to properly analyze features and functionality means that relationships between content types and interactions are not modeled consistently. Instead of a stable, consistent system, we end up with a mash-up of implementation approaches, each of which worked for the part of the system that was being developed at the time. So we lose consistency.

Maybe the best solution is to have project sponsors who understand that any content system is really about communications and that authoring integrity is fundamental to good communication. Such sponsors may understand that the foundation needs to be laid first – that you need a blueprint before you build a house.

Agile design is not always bad; it can be appropriate for some undertakings (a start-up, where a pivot is expected and where every premise of the system is up for reconsideration, or an evolutionary project, adding to an existing, stable system with consistent paradigms). Developing a CMS with a good author experience is not one of these. For this, we need the full backlog of meaningful stories, consistent paradigms, and a map of dependencies so the developers can put the foundation down first.

# CHAPTER 9
# *In Conclusion*

The digital realm is no longer an afterthought to communication. Nor is digital restricted to the web site. Digital media are now the trunk routes through which all communication flows and is managed, whether the output is on a screen or some non-digital format.

This is a fundamental shift in the inner workings of business, one that many organizations are struggling with or ignoring in the hope that it does not apply to them.

## The evolution of content management

Demands from the consumers of published information are forcing organizations to change the way they think about, and handle, content. These changes require new patterns and processes. Author experience can provide the structure for content management systems to evolve in ways that enable the people responsible for content to stay in control while coping with the needs for adaptive content delivery.

### *Integrated information flow*

The world is experiencing a shift in the perceived functional purposes of content. What, in the past, was a series of specific scopes for content – internal, marketing, support, investor, etc. – has already seen an overlapping of function and audience boundaries. This shift will continue until they merge and the distinctions are all but erased.

Figure 9.1 – Evolution of content purposes

This does not mean that customers will see confidential internal material or that competitors will have access to a company's plans while they are still in development. It means the lines will blur between information purposes. Within the scope of what one has access to, information will be integrated, without artificial boundaries.

## Distributed publication

Just as the boundaries between content sources blur, so do the channels through which content is consumed. Whatever we create needs to be accessible to all appropriate audiences, however they choose to engage with us.

And that is the crux of the matter: your audience chooses how they access content. They do not care how you think your content should be presented. They do not go out of their way to access it through your preferred channel.

Soon, if your content is not adaptive, you will lose access to the majority of your audience.

## Give me semantics

If you want to reuse content for multiple audiences and if you need your content to be easily and programmatically adapted to the myriad platforms and channels currently in use (and countless more to come), then that content must be structured semantically. We cannot afford the old model of human-driven content repurposing.[1] The sources are too complex and the output channels too numerous to rely upon manual processes.

Only semantically rich content, garnished with metadata, can possibly feed the information needs of all our audiences across all the channels they choose to use.

## Tools for the job

Managing complex, semantically rich content requires new tools. These tools must be able to handle semantic, structured content and translate that structure to interfaces that make sense to authors. Content responsibility equates to knowledge responsibility, and we cannot expect every subject matter expert in every field to become adept at managing the semantic descriptors that give their content structure. They need to have authoring interfaces that transparently translate their thinking – their domain models – to the system's internal structures.

These tools must be simple to use, dedicated in function. Technology must do what it is supposed to: facilitate human tasks, carry out the burdensome computations, and hide the complexity.

---

[1] Even in the classical digital realm of web sites, the dual-management of a desktop and a mobile site has proven too much for most of the organizations that have tried.

> **A purpose-built tool**
>
> A brilliant example of a content management tool built for a specific purpose and designed to integrate with its users' needs and workflows is Scoop, the New York Times CMS. According to Luke Vnenchak, "Scoop was initially designed and developed in 2008 *in close partnership with the newsroom.*" (Emphasis added.)[2]
>
> Scoop is modeled on the needs of journalists. It provides the features and functionality requested by the people writing and editing stories for publication in online (and offline) news media, including the ancillary roles of asset management. It models not only the basic tasks of writing the individual piece, but also the output-aggregating content type (putting together the print layout). Its concept of workflow is based on the way evolving news stories flow through the editorial process. It has sufficient metadata logic – including the ability to distinguish between the same word used in different contexts (e.g. Georgia: a US state and a Eurasian country) – that it can intelligently suggest related content.
>
> Simply, Scoop was designed and built to facilitate the process of managing news content.

## Enter author experience

The evolution of content management will not occur on its own. For tools to be developed that are suited to the needs of content managers, the specifications must be developed by people who understand the needs of the organization, how content evolves and is used, and the underlying technical principles of the implementation.

Author experience is the one field that concentrates on the management and flow of content from its inception, through a user-appropriate input interface, into an adaptive content store that can serve all channels.

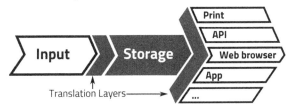

Figure 9.2 – The author experience view of communication

---

[2] http://open.blogs.nytimes.com/2014/06/17/scoop-a-glimpse-into-the-nytimes-cms/

With time, applied author experience can see the issues identified in Chapter 4 fade to bad memories. We will see fit-for-purpose language, associative structured content, and everything else described in Chapter 5, becoming standard and expected aspects of every CMS.

Making content management fit for purpose may start with individual implementations, customizations of existing platforms to do what organizations expect of their content. But eventually, the flexibility will seep into the wider field. Author experience will be an expected part of every implementation, and off-the-shelf products will support domain-modeled customization.

## Moving forward

### *A dedicated field*

The skillset required to provide a comprehensive author experience consultancy is not trivial. It encompasses high-level business-focused requirements gathering, analysis, and domain modeling as well as the technical skills of information modeling and implementation.

This is not a set of integrated skills that can be developed and honed as an adjunct to a more general practice. Just as UX research, interaction design, visual design, and content creation have become specialized fields, author experience is becoming a dedicated discipline aimed at providing appropriate, domain-modeled functionality within content management environments.

If you want a content management solution that meets your communication needs and is not a CMmess, the key is having an author experience consultant on board to analyze requirements and define the content workflow based on your organization's domain model.

As CMSes evolve to enable straightforward configuration of their author interfaces, the need for author experience specialists will grow. Implementations may become more streamlined, but that will not reduce the need for skilled individuals who understand communication needs and the business processes surrounding them and who can translate those needs into a model that aligns content management with the business.

## For clients: you are not a nail

Any consideration of digital communication starts with two constants:

- You have information you need to package and deliver to your audience.
- A set of technologies will be used to deliver that information.

These two constant end-points are common to everyone. Unfortunately, many implementers operate on the premise that the path between these two end-points can be standardized, that one solution – their solution – is suitable for all.

In reality, the most suitable path between these two known quantities is unique and depends on the complexity of your organization. To plot the correct path for your organization, you need to understand and map the processes that hold sway within your domain.

If you want your communications to be meaningful and effective, you need to map out how they work internally – the flow and processes. Then you need to determine what technical solution best serves you.

The best solution may leave some processes manual, handling only the delivery end-point; it could be one that helps you manage your information across all its uses, throughout its lifecycle; or it could be something in between.

There may be an off-the-shelf system that matches your requirements. But to know this, you must first know your needs. Do not be the nail to a CMS vendor or implementer's hammer.

## For implementers / vendors: your duty of care

It is understandable that as an implementer or vendor of CMSes, you believe yours is perfect. It may well be perfect, for the specific function you envisaged. That does not mean it is perfect, suitable, or in any way good enough for a particular client.

If you are providing a content management solution, the aim is not to sell your product or service (much as you may be driven by profit); your aim must be to satisfy the client's needs. To do this, you must understand what those needs are, not the checklist of features they presented to you in the RFP. And consequently, your proposed solution must adapt itself to the client's needs and its paradigms for defining, managing, and using content.

To paraphrase a discussion I had recently with Jeff Eaton[3] about questions to put to a conference panel on enabling content agility from a content strategist's perspective, "Tell me why your CMS is better than notepad." There is, obviously, only one correct answer: "First, tell me what you want to achieve."

## *For practitioners: the long road ahead*

Much as I would like to say that we are very nearly there – that soon we will have fit-for-purpose content management systems that are smart enough to proactively help us manage content – we still have quite a way to go.

The technology itself will evolve in fits and starts with new innovations followed by everyone playing catch-up. We will gradually see more features and functionality aimed at making the management aspect of the content lifecycle better. But this is not going to happen overnight; it will take several years.

And once the technology catches up, awareness that content management can and must be domain modeled will seep into the general consciousness, and the demand for those who specialize in author experience will grow. Then, we will see many years of growth and refinement within the field: new design patterns, reusable approaches, best practices.

The future of author experience is bright.

---

[3] Jeff (@eaton on Twitter) is a Senior Digital Strategist at Lullabot, an interactive strategy, design, and development company.

# Glossary

**adaptive content**
    Content that is designed to adapt to the needs of the customer, not just cosmetically, but also in substance and in capability. Adaptive content automatically responds to the screen size and orientation of any device, but goes further by displaying relevant content that takes full advantage of the capabilities of the device being used.

**axis**
    A metadata axis identifies a category of metadata. For example, a person involved in a movie might be identified in the director or actor axis. Metadata axes may limit the type of categorization values they can accept.

**content**
    Any text, image, video, decoration, or user-consumable elements that contribute to comprehension.

**content audit**
    The process and result of conducting a quantitative study of a content inventory.

**content management system**
    A software application that supports information capture, editorial, governance, and publishing processes with tools such as workflow, access control, versioning, search, and collaboration.

**DITA**
    DITA (Darwin Information Typing Architecture) is an XML data model for authoring and publishing. DITA was originally developed by IBM and in 2005 was approved as an OASIS (Organization for the Advancement of Structured Information Standards) standard. DITA provides structure mechanisms and extensibility.

**domain model**
    A domain model is a conceptual model of all the topics related to a specific subject. It describes the entities, their attributes, and particularly the relationships between them. It is how those involved in the subject think about the elements it comprises.

**latent semantic indexing**
    Latent semantic indexing (LSI) is an indexing and retrieval method that identifies patterns in the relationships between terms and concepts contained in unstructured text. LSI is based on the principle that words used in the same contexts tend to have similar meanings.

**metadata**
    Attributes of content you can use to structure, semantically define, and target content.

**multi-axis profiling**
    Multi-axis profiling is the process of categorizing an individual based on a variety of known or assumed social, personal, preferential, and behavioral traits and identifying a segmentation profile based on the combination of attributes. Common attributes include age, gender, country, economic status, purchase history, etc.

**page parity**
    Page parity identifies the case where there is a one-to-one relationship between pages in different environments.

**rich text**
    Rich text is copy that includes presentational elements such as fonts, colors, text sizes, styling, or even other media (images, video, etc.) within a block of text.

**tag**
    A tag is a simple, human-readable, text metadata value. The tags metadata axis provides a general category. Tags are the most common type of free-form metadata.

**technical debt**
    Technical debt is a term used in software development to describe the consequences of poor system design or coding practices. The debt is the effort required to bring the code up to expected functional, performance, and maintainability standards.

# *About the Author*

Rick Yagodich (@think_info on Twitter) has been working with the web since 1995. Right after creating his first HTML page, he asked "What is the business purpose of this site?" That question has driven his interactions with web technology ever since.

He founded his company, Excolo,[1] as a vehicle to offer web consultancy in 1997. Since then, he has served clients in a variety of industries: from financial services to retail; from telecommunications to business intelligence.

Rick has experience with most aspects of web technology, from front-end design to back-end development. It is from this breadth of experience that he determined a vital part of the industry was severely underserved: the systems used to manage content are not designed for the needs and processes of those who use them. Excolo now focuses on that niche of the web consultancy market: Author Experience.

---

[1] http://excolo.com

# Index

## A

abstracted semantics, 22
accessibility, content, 62–66
acting, as communication, 23
adaptive content
　described, 43
　previewing, 79–80
adaptive content delivery, 4, 75–77
adaptive presentation, 75–77
agile methodology, 145–146
algorithmic association, 86
amateur communicators, 22
analysis, technical business, 96–98
analytics, integrating, 108
appropriateness, author context and, 11
approvals
　multi-person, 34
　problems with, 36
　workflow, 33–37, 82
archives, ownership of, 39
articles, structure of news, 99–101
associations
　algorithmic, 86
　auditing content, 101–102
　content, 27–28, 61–62, 82–83
　created by metadata, 62
　dynamic, 71
associative references, 69
associative structured content, 66–74
attributes, managing, 58–60
audits
　author experience, 93
　content, 93
　existing-system, 94
　forming a team, 95
　linking and, 101–102
　new-system, 94–95
　prioritizing results, 111–112
　tools, 95
author environment, defined, 7

author experience
　as a field, 141–146, 150
　business value of, 13–15
　challenges to, 21
　defined, 3, 10–11
　hierarchy of needs, 53
　measuring quality, 112–115
　pillars of, 13
　vs. user experience, 19–20
author experience audit, 93
　basic questions, 96
　goals of, 93–95
　technical business analysis, 96–98
　tools, 95
author experience design, 101–109
author experience design patterns (*see* design patterns)
authoring
　abstractness and complexity in, 12
　hub-and-spoke, 107
authors
　adapting interfaces to, 89–91
　amateurs vs. professionals, 22
　defined, 7
　resistance to change from, 143–144
　training, 12, 113
axes
　content-specific, 126
　dependent, 48–50
　filtering and searching on, 65–66
　in content consoles, 125–127
　vs. tags, 46

## B

Bailie, Rahel
　content validation model, 67
baking, as analogy for reuse, 68–69
BCD (business-centered design), 142
behavior
　coupling, 32
　defined, 30

# 158 Index

previewing, 80–81
Bloomstein, Margot, 107
business analysis, technical, 96–98
business process evolution, 115
business process governance, 100–101
business value, author experience as, 13–15
business-centered design (BCD), 142

## C

categorization, 46
change, resistance to, 143–144
clustering, content, 88
CMmess, 13–14
CMS (*see* content management system)
communication
    analyzing, 42
    cost of mistakes in, 14–15
    flow of, 3–4
    forms of, 23
    goals of, 23–25
    knowing how to communicate, 21–23
    parts of, 3–4
    process, 3–4
    translation layer, 4
company publications, governance challenges, 100
complexity
    compared with complicatedness, 44
    content, 44–45
    modeling, 98–101
conceptual descriptor metadata, management of, 122
conditional metadata, 49
conditional text, 32
consistency
    in interfaces, 102
    structural, 58–60, 103
consoles, content, 124–127
constraints, structural, 67
content
    adaptive, 43
    as static message, 24
    associations, 27–28, 82–83
    associations, auditing, 101–102
    associative structured, 66–74
    compared with information, 10
    complexity, 44–45
    defined, 30
    evolution of, 147
    filters (*see* filters)
    granular (*see* granular content)
    managing large amounts, 125
    medium as, 24
    non-displayable, 84
    ownership, 37–40
    page-based, problems with, 42
    portable, 129
    reuse, 68–69, 148
    reuse, analytics and, 108
    risks, 28
    self-aware, 86–89
    structured, 67
    sub-structured, 102–103
    syndicated, 40
    user context and, 77–79
    value in, 22–23
content accessibility, 62–66
content aggregation tool, 106
content audit, compared with AX audit, 93
content consoles, 124–127
content coupling, 30–33, 66
content delivery, adaptive, 44
content design, governance and, 98–101
content grouping, contextual, 88
content identifiers, 27–28
content lifecycle, 83
content management
    agile methodology and, 145–146
    effectiveness of workflows, 33–37
    evolution of, 147–149
    optimal model for, 94
    per device, 33
    role of, 8–10
    tools, 81–86, 148–149
    web, 8
content management system (CMS)
    analytics integration, 108
    dependency awareness in, 87–88
    flexibility, 29–30
    implementation process, 17–19
    managing limitations, 109
    managing metadata, 46–50
    problems with, 26–27
    purchasing, 15–19
    purpose of, 9–10
    reasons for, 9
    training people to use, 12
    types of, 13
    upgrades, 28–29, 114–115
    vendors, 9

workflows in, 33–37
content models (*see* models)
content storage (*see* storage)
content type identification, 98–101, 126
content validation model, 67
context
    coupling, 32
    defined, 30
    input, 78
    user, adapting content to, 77–79
contextual content grouping, 88
contextual descriptors, 119–120
contextual presentation, 84
contextual presentation models, 77
contextually appropriate functionality, 11
COPE (Create Once, Publish Everywhere), 105–106
Copeland, Todd, on mandatory and optional fields, 105
coupling
    behavior, 32
    content, 66
    context, 32
    presentation, 31
Create Once, Publish Everywhere (COPE), 105–106
custom vs. open source platforms, 14
customization, cost of, 28

## D

Darwin Information Typing Architecture (DITA), 32
date, selecting content by, 132–134
delivery, contextually dependent, 24
dependencies
    managing, 136
    prioritizing, 112
dependency awareness, 87–88
design
    agile, 145–146
    author experience, 101–109
    business-centered, 142
    User-Centered, 142
design patterns
    content consoles, 124–127
    integrated label use, 120–121
    label structure deployment, 122–124
    narrative, 127–128
    referrer/referee links, 134–136
    selection by date, 132–134
    stored or searched references, 136–137

    UI label management, 117–120
    WYSISMUC, 129–131
DITA, 32
Donaldson, Krista, on customers and products, 20
dynamic associations, 71
dynamic multi-axis content filtering, 65–66, 126

## E

Eaton, Jeff, on CMS implementation, 152
Eckman, John, on the CMS Myth, 33
editors, WYSIMUC, 130–131
embedded help, consistency in, 102
emulation layer in CMS, 79
error messages, consistency in, 102

## F

filtered types, 64
filters
    content, 71
    displaying results, 135–136
    multi-axis content, 65–66, 84–85
fit-for-purpose language, 54–62
fit-for-purpose system, 113
flexibility in content management systems, 29–30
flexible trigger, 90
flow
    integrated information, 147
    narrative, 71–72, 127–128
    process, 97
folders, as structural model, 27
forms
    as interfaces, 105–109
    visual presentation, 106
functionality, excess, 94

## G

GatherContent, content aggregation tool, 106
Ghoshal, Sumantra, on best practices, 28
Gladwell, Malcolm, on reduced information sets, 103
Goldman, Lee, on reduced information sets, 104
governance
    amount of work required, 101
    business process, 100–101

content design and, 98–101
dependencies and, 87
effects of semantic referencing on, 71
ownership and, 37–40
granular content
    challenges accessing, 62
    described, 43
    how much is too much, 103–104
    output-based navigation and, 63
Gratton, Lynda, on best practices, 28

## H

hierarchy of author experience needs
    associative structured content, 66–74
    content accessibility, 62–66
    content management tools, 81–86
    fit-for-purpose language, 54–62
    rules-based presentation, 75–81
    self-aware content, 86–89
hierarchy of needs, Maslow's, 53
hierarchy, information, 72–74
hub-and-spoke model
    as authoring paradigm, 107
    converting pages into, 81
    for smaller sets of content, 103
    relevance of, 71

## I

identifiers, content, 27–28
images
    configuring attributes for, 59
    content management of, 74
information
    as currency, 80
    compared with content, 10
    flow, integrated, 147
    proximity of, 74
information architecture, 109–110
information hierarchy, 72–74
information sets, reduced, 103–104
information technology (IT), role in purchasing, 15
input
    in communication, 4
    presentation, 78
integrated labels, 120–121
interfaces
    adaptive, 89–91
    forms-based, 105–109
    learning, 89–91
    measuring value of, 113–114
    predictive, 90–91
International Date Line, 132
IT (information technology), role in purchasing, 15
iterative content association, 82–83

## J

Johns, Max, on risk management, 82

## L

labels (*see* user interface labels)
language
    consistency of, 57
    disconnects in, 55
    fit-for-purpose, 54–62
latent semantic indexing, 90
layers
    translation, 4, 29
    translation, resistance to, 142–143
    translation, upgrades and, 114
learning systems, 89–91
lifecycle, content, 10, 83
linking paradigms, 102
links
    associative, 69
    design pattern, 134–136
    one-to-one hard, 61
    vs. associations, 61
localization, narrative flexibility and, 128

## M

management
    attribute, 58–60
    logic, consistency of, 58–60
    micro-copy, 119–120
    perception of tech comm, 22–23
mandatory content, managing, 105
markup
    presentational vs. semantic, 72–74
    structured, 130
Maslow, Abraham, hierarchy of needs, 53
Means, Garann, on developer UI, 26
media, dependencies and, 87
mental model, 40–45
meta-metadata, 48–50
metadata
    auto-generated, 90
    conditional, 49

contextual descriptor, 122
filtering based on, 84–85
grouping content with, 89
managing, 58–60
mandatory and optional, 47
multi-axis, 46
overload, 104
problems with, 46–47
retaining excess, 49
search based on, 84
types of, 47
methodology, agile, 145–146
micro-copy
consistency in, 102
contextual descriptors for, 119
described, 43
hard-coded, problems with, 118
importance of, 118
managing, 119–120
ownership of, 121
screenshots and, 121
minimum viable distinction, 104
mistakes, cost of communication, 14–15
mobile
hub-and-spoke model, 71
redesigning for, 31
models
complexity, 98–101
contextual presentation, 77
disconnects in, 55
filtered types, 64
hub-and-spoke, 71, 81, 103
internal content, 26
mental, 40–45
optimal content management, 94
storage access, 63, 102
structural, folders as, 27
tree, 63–64
more-content dilemma, 25
multi-axis content filtering, 65–66
multi-axis metadata, 46
multi-axis profiling, 25
multi-channel adaptive output, 4

## N

narrative flow, 71–72, 127–128
navigation, output-based, 63
New York Times
correction of archived article by, 39
Scoop, 149
news articles, structure, 99–101

## O

open source vs. custom platforms, 14
output
in communication, 4
preview, 79
simulated, 79
output-based navigation, 63
ownership
archives, 39
content, 37–40
evolution of, 38
silos and, 144–145

## P

page parity, 42, 63
pages
as mental model, 40
as structural model, 27
problems with, 41–42
references to, 83–84
paradigms
developer vs. author, 28
technologist's, 25
portable content, 129
predictive systems, 90–91
presentation
adaptive, 75–77
contextual, 84
defined, 30
rules-based, 75–81
presentation coupling, 31
presentational markup, 72–74
previewing
adaptive, 79–80
behavioral, 80–81
in-CMS vs. dedicated, 80
process flow, 97
professional communicators, 22
profiling, multi-axis, 25
proximity of information, 74
purchasing
content management tools, 15–19
pitfalls in, 16

## Q

quality, measuring author experience, 112–115

## R

recursion, 44
references, 134
   (*see also* associations)
   (*see also* links)
   associative, 69
   caching, 136
   dependencies, 136
   design pattern, 134–136
   inbound, 136
   page, 83–84
   rules-based, 86
   storing inbound, 137
Reiss, Eric, on user experience, 19
relationships, semantic, 69–70
release, staggered, 81
repurposed content, 68–69
requirements, prioritizing, 112
responsive web design (RWD), 75–77
reuse
   analytics and, 108
   content, 68–69, 72–74, 148
risk management, 82
roles, author, 7
rules-based presentation, 75–81
rules-based references, 86
RWD (responsive web design), 75–77

## S

Scoop, New York Times, 149
search
   by date, challenges, 133–134
   in content consoles, 125–127
   multi-axis, 65–66, 84–85
self-aware content, 86–89
semantic markup, 72–74
semantic relationships, 69–70
semantic structure, 148
sequence, narrative flow and, 71–72
silos, organizational, 144–145
slug, as a unique identifier, 56
speaking, as communication, 23
staggered release, 81
storage
   content, 8
   content management models, 102
   dynamically adaptable, 45
   in communication, 4
storage, as a part of the communication process, 3

structure
   complexity in, 44–45
   consistency in, 58–60
   designing content, 98–100
   folders as, 27
   lack of, 31
   narrative, 127–128
   page, 41–42
   pages as, 27
   semantic, 148
structured content
   associative, 66–74
   defined, 68
   markup, 130
sub-structured content, 102–103
syndicated content, 40
synonyms
   equality of, 55–56
   problems managing, 126–127
   unnecessary use of, 57

## T

tags vs. axes, 46
team, audit, 95
technical business analysis, 96–98
technology, role in content management, 26
terminology
   consistency of, 54, 57
   obscure, 56
   synonyms, 55–56
Thompson, Matt, on training authors, 12
time stamps, 133–134
time zones, 132
titles
   job, in author experience, 141
   varieties of content, 56
tools
   author experience audit, 95
   content management, 81–86, 148–149
   importance of, 15
   importance of quality, 22
trans-algorithmic dependency awareness, 88
transactions, behavioral previewing of, 80
translation layer, 4
   expense of, 28
   interfaces for building, 29
   optimizing communication with, 4
   resistance to, 142–143
   upgrades and, 114

tree model, 63–64
tree view, selecting content using, 85
triggers, flexible, 85–86
types, filtered, 64

## U

UCD (user-centered design), 142
upgrades
    business process, 115
    content management system, 28–29, 114–115
Urbina, Noz
    on adaptive content, 77
    on conditional text in DITA, 32
user experience
    defined by Eric Reiss, 19
    vs. author experience, 3, 11, 19–20
user interface labels
    authoring, 120
    consistency in, 102
    deployment of, 122–124
    integrated, 120–121
    managing, 117–120
    screenshots and, 121
user-centered design (UCD), 142

## V

vendors
    content management system, 9
    role in author experience, 143, 151–152
Vnenchak, Luke, on NYT's Scoop, 149

## W

web content management, 8
    (*see also* content management)
web design, responsive, 75–77
What You See Is Structurally Marked-Up Content, 129–131
Wikipedia, associative linking on, 70
workflow
    approvals, 82
    as a review process, 33
    as content management challenge, 33–37
    dependencies and, 37
writing, as communication, 23
WYSISMUC, 129–131
WYSIWYG, 129
WYSIWYM, 129

## Z

zones, narrative, 127

# Colophon

## About the Book

This book was authored, edited, and indexed in a Confluence wiki. Contents were exported to DocBook using the Scroll DocBook Exporter from K15t Software. The book was then generated from that output using the DocBook XML stylesheets with XML Press customizations and, for the print edition, the RenderX XEP formatter.

## About the Content Wrangler Content Strategy Book Series

The Content Wrangler Content Strategy Book Series from XML Press provides content professionals with a road map for success. Each volume provides practical advice, best practices, and lessons learned from the most knowledgeable content strategists in the world. Visit the companion website for more information contentstrategybooks.com.

## About XML Press

XML Press (xmlpress.net) was founded in 2008 to publish content that helps technical communicators be more effective. Our publications support managers, social media practitioners, technical communicators, and content strategists and the engineers who support their efforts.

Our publications are available through most retailers, and discounted pricing is available for volume purchases for educational or promotional use. For more information, send email to orders@xmlpress.net or call us at (970) 231-3624.

# The Content Wrangler
# Content Strategy Book Series

The Content Wrangler Content Strategy Book Series from XML Press provides content professionals with a road map for success. Each volume provides practical advice, best practices, and lessons learned from the most knowledgeable content strategists in the world.

*The Language of Content Strategy*

Scott Abel and Rahel Anne Bailie

*Available Now*

Print: $19.95
eBook: $16.95

*Content Audits and Inventories: A Handbook*

Paula Ladenburg Land

*Available Now*

Print: $24.95
eBook: $19.95

*The Language of Content Strategy* is the gateway to a language that describes the world of content strategy. With fifty-two contributors, all known for their depth of knowledge, this set of terms forms the core of an emerging profession and, as a result, helps shape the profession. The terminology spans a range of competencies with the broad area of content strategy.

Successful content strategy projects start with knowing the quantity, type, and quality of existing assets. Paula Land's new book, *Content Audits and Inventories: A Handbook*, shows you how to begin with an automated inventory, scope and plan an audit, evaluate content against business and user goals, and move forward with a set of useful, actionable insights.

*Enterprise Content Strategy: A Project Guide*

Kevin P. Nichols

*Available Summer, 2014*

Print: $24.95
eBook: $19.95

*Global Content Strategy: A Primer*

Val Swisher

*Available October, 2014*

Print: $19.95
eBook: $16.95

Kevin P. Nichols' *Enterprise Content Strategy: A Project Guide* outlines best practices for conducting and executing content strategy projects. His book is a step-by-step guide to building an enterprise content strategy for your organization.

Nearly every organization needs to serve customers around the world. This book describes how to build a global content strategy that addresses analysis, planning, development, delivery, and consumption of global content that will serve customers wherever they are.

ContentStrategyBooks.com
XMLPress.net

CPSIA information can be obtained at www.ICGtesting.com
Printed in the USA
BVOW11s1540141014

370715BV00001B/1/P